CATACLYSMS OF THE EARTH

The HAB Theory Reloaded

Hugh Auchincloss Brown

Adventures Unlimited Press

And Nature, the old nurse, took
The child upon her knee,
Saying "Here is a story book
Thy Father hath written for thee."

"Come wander with me," she said,
"Into regions yet untrod,
And read what is still unread
In the manuscripts of God."
—Longfellow

CATACLYSMS OF THE EARTH

The HAB Theory Reloaded

Cataclysms of the Earth

ISBN 978-1-939149-70-1

Copyright 1967

Hugh Auchincloss Brown

This edition © 2016

All rights reserved.

Published by Adventures Unlimited Press
One Adventure Place
Kempton, Illinois 60946 USA

www.adventuresunlimitedpress.com

Printed in the
United States of America

CATACLYSMS OF THE EARTH

The HAB Theory Reloaded

There have been, and there will be again, many destructions of mankind. When civilization is destroyed by natural calamities, then you have to begin all over again as children.
—High Priest to Solon in Plato's *Dialogues*

FOREWORD
BY DAVID HATCHER CHILDRESS

Hugh Auchincloss Brown's 1967 book, *Cataclysms of the Earth*, was to be become a classic by the 1970s where whole communities of survivalists, called "preppers" today, read the book along with *Mother Earth News* waiting for the coming catastrophe. The disaster movie *2012* featured a cataclysmic pole shift as described by Brown in his book.

In fact, Brown had spent a lifetime promoting his theory of ice caps and crustal slippage that caused catastrophes in the past, and would in the future as well. Brown was 91 when his book was finally published in hardback by a New York vanity publisher. Brown reportedly paid for the printing. It was received well by the cataclysmic community, including those who were reading Immanuel Velikovsky's books such as *Earth in Upheaval, Worlds in Collision, Mankind in Amnesia* and others. Velikovsky's books were wildly popular in the late 60s and early 70s, spawning all sorts of fringe science books such as *Chariots of the Gods, The Bermuda Triangle* and *The Twelfth Planet*.

Hugh Auchincloss Brown was born on December 23, 1879 (he died November 19, 1975) and graduated from Columbia University in 1900. He was an electrical engineer best known for advancing a theory of catastrophic pole shift. Brown is said to have had a long and distinguished career as an engineer, inventor and businessman. He spent most of his life searching for scientific evidence that would prove his theory of pole shifts wrong. Everything he found reinforced it.

Brown's book became the basis for a popular 1970s mass market paperback called *The HAB Theory* by Allan W. Eckert. The book was first published in hardback by Little, Brown & Co. and then published by New York's Popular Library in 1977 at which time it became a cult favorite and went through several printings. In those days, before cable television or the Internet, pulp fiction books bought in bookstores, supermarkets and drugstores were a major part of the entertainment industry.

In Eckert's book, the fictional character is named Herbert Alan Boardman (HAB) and is based on Hugh Auchincloss Brown. The theory in Eckert's book is based on Brown's theories as elucidated in *Cataclysms of the Earth*.

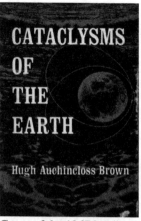

Cover of the 1967 hardback.

Eckert has said that he wrote *The HAB Theory* in order to use his novel as a vehicle for public awareness of the possible impending catastrophe as envisioned by Brown. Eckert said he wanted to point out the scientific self-defeatism of intense specialization in various scientific fields, a point also made by Brown.

Both Eckert and Brown in their books show grave concern about scientific tunnel vision, saying that while it may be good for one's career, it can also be dangerous. If an individual scientist's specific field consumes all of his or her attention, then that person may not look at it sufficiently in its relationship to other fields of scientific endeavor. Both Brown and Eckert were concerned that very few scientists seem to grasp the various larger pictures that can be drawn from the huge fields of science.

Brown's book gave reasonable answers to such questions as why mammoths would be flash-frozen in the Arctic and how the mysterious process of fossilization works. Brown claimed that massive accumulation of ice at the poles caused recurring tipping of the axis in cycles of approximately 4,000–7,500 years. Brown argued that because the earth wobbles on the axis and the crust slides on the mantle, a shift was demonstrably imminent, and suggested the use of nuclear explosions to break up the ice to forestall catastrophe.

Asked Brown's early publishers in their cover copy: "By what process were entire tropical forests and volatile raindrops instantly arrested and fossilized? What sensible explanation can be given to the fact that each stratum of the earth's composition seems formed by a different climate?"

Brown's book quickly fell out of print, but interest in it never declined and it is now back in this deluxe edition. Brown's grandson, Hugh Auchincloss Brown III, recalled in an interview in 2003:

> Granddad played college baseball for Columbia University, including their annual exhibition games against the New York Giants. Sixty years later, Grandad still loved to tell the tale of the times he got a hit off Giant's ace pitcher Amos Russie who was a big star in the 1890s. The Giants were New York's National League team from the late 1800s until the mid 1950s when they moved to San Francisco.
>
> Granddad always dressed in a suit, walked and sat with a straight posture, and was a wonderful conversationalist. He drove his car into his nineties and walked a mile every day until his death in 1975. His two sons have passed away, but he has three grandsons and two granddaughters still living.

It is a good thing that *Cataclysms of the Earth* is back in print. And while some of Brown's theories may seem dated today, his book remains as important in this era as it was in 1967. In my mind, it is time to take another look at *Cataclysms of the Earth*.

—David Hatcher Childress, September 2016.

Contents

	Introduction	1

PART ONE

I	Evidence of Careenings of the Globe	13
II	Mechanics of the Great Deluge	119
III	Man: the Past, the Present, and the Future	136

PART TWO

I	Sir Isaac Newton on Gravitation	159
II	Physical Evidence Refuting the Universal Mutual Attraction of Masses Theory	167
III	The Growth of the Theory of Gravitation Repulsion	173
IV	Dynamic Electricity	177
V	Static Electricity	192
VI	Research Projects	210

PART THREE

I	Origin of the Earth's Materials	217
II	Pre-Historic Forest Floors	226
III	The Creation of the Oceans	260
IV	A Cosmic Cycle in Time	269
	Index	277

Hugh Auchincloss Brown holding his doomsday demonstration device held on a string, circa 1967. He was 91 at the time.

INTRODUCTION

We are all aware that our earth is constantly changing. By the roadside or in the mountains we notice the layers of rocks, or strata, displaced from their original horizontal plane, and recognize the evidence of the shifting of the earth's so-called crust. In the cross sections of canyons we note the succession of layers—the permanent markings made by rock formation in earlier epochs. We uncover animal and plant fossils and by their depth and placement can construct a chronology of prehistoric life. The movement of the rivers carving out new beds and the pounding of the oceans on their shores all remind us of the endless motion of the land and seas.

These, however, are but the surface phenomena of the evolution of our globe. Rarely do we speculate on the forces constantly reforming the planet as a whole; rarely do we speculate on the nature and the effects of major upheavals and cataclysms that, on a larger scale, have characterized the history of the earth. We are all aware of the great stone book that is our earth, but rarely do we read its pages. The origin of our planet remains a mystery; of its life history, inscribed upon itself, only minutiae are fully known to man.

The successive layers in the earth's crust reveal that the earth has not only undergone vast surface changes, such as the stratification of layers of rock and the formation of mountain ranges and river beds. On a much grander scale, the evidence suggests that the earth's axis of rotation has repeatedly changed its direction; during its journeys around the sun the earth has been rolling sideways as it spins in daily rotation, and shifting to a new Axis of Figure, i.e., the axis defining the geographic poles.

Such an axial shift can best be understood when considered

in relation to the earth's rotation about the Axis of Spin—the true axis of rotation in respect to the stars and irrespective of the surface terrain of the earth. The Axis of Spin is the imaginary line in space about which the earth rotates; it is constant in relation to the earth itself, but changes its position in space as described later. The Axis of Figure is an imaginary line through the center of the earth; it is defined by the arbitrary designation of the geographic poles, wherever they may occur during any epoch, and, being relative to the geography of the earth's surface, it changes whenever the earth shifts about the Axis of Spin.

FIG. 1. Deviation between true axis and Axis of Figure.

Although the earth has continued to rotate on the Axis of Spin that passes through Polaris in the northern sky and the Southern Cross in the southern sky, it has in the course of time turned sideways or "careened" in its spinning; the geographic positions of the poles, the Equator, and the various zones of the globe, have been redefined in respect to the surface terrain. During such careening the Equator tilts, and the poles, defining the Axis of Figure, rotate away from the Axis of Spin. All areas

Introduction

of the globe, except those two points directly on the Axis of Careen, are altered in relation to the Axis of Spin, and hence in relation to their previous climates. A land mass or ocean, once in an equatorial zone, is shifted towards a polar position; the poles shift around towards the equatorial zones. The previous Axis of Figure is then replaced by a new one, determined by the new geography of the earth. At the beginning of each epoch, the Axis of Figure and the Axis of Spin coincide.

Fossils of animal and plant life from previous epochs serve as telltale indicators of the climate prevailing when the strata in which they are found formed the earth's surface. From fossils and other evidence we are able to document the displacement of materials from their native climates to areas of new and often alien climates and to designate the amount of displacement and the date of earlier careenings of the globe. From the condition of the evidence, we are able to suggest the cataclysmic forces that took effect at such time.

This evidence, however, provides us only with the symptoms of the forces at work on our planet. To fully understand *why* the earth has come to be as we know it, we must also analyse *how* it came to be that way.

We must construct a theory, consistent with all the data that we can collect, which will explain how such careenings have come about. And we must test this hypothesis by applying it to the earth as we presently know it.

Here again, all the evidence mentioned so far will serve as our guide. Indeed, the discovery of the recurrent careening of our globe resulted from efforts directed at solving the mystery of prehistoric animal life.

The value of any branch of science, even such a seemingly "backward" looking one as geology, lies not only in the knowledge we may gain about the past and the changes and developments that have occurred. For science to be of value in the present, for it to transcend the mere recording of the past and present, it must offer theories that are applicable to the future.

Years of investigation and research, coupled with resolution and courage to follow wherever truth might lead, have established the certainty of a future world cataclysm during which most of the earth's population will be destroyed in the same manner as the mammoths of prehistoric times were destroyed. Such an event has occurred each time that one or two polar ice caps grew to maturity; a recurrent event in global history it is clearly written in the rocks of a very old earth.

The earth is approximately 4½ billion years old. Human beings have been living on it for at least 500,000 years and perhaps even one million years. To appreciate the immensity of these figures, one might imagine the age of the earth represented by the period of about one week; the duration of our own epoch, 7,000 years, is then but one second! By a similar analogy, men have lived on an earth that is one week old for just two minutes. It is evident that our own epoch is but a very short and insignificant period in the life of our planet and our species.

In past epochs there have been ice caps at one or both of the geographical poles. The heat of the sun caused these ice caps to grow larger. As the sun heats the air of the hemisphere, the heated air expands, becomes lighter, and rises. The updrafts are greatest in the tropics. As the earth is virtually spherical, the currents of warm air converge at the poles. Meeting head-on from every direction, they create areas of air pressure, become colder and heavier, turn downward, reversing the direction of their flow, and pour back toward the Equator from the polar centers with high velocities. Thus, there is a continuous circulation of rising humid warm air journeying poleward and a down draft of cold dehumidified air returning from the poles at low or ground altitudes. Air acts like a sponge. When warm, it absorbs water; when cold, it cannot hold much water, and in cooling releases any surplus moisture to fall as rain or snow.

Most of the snow that falls in the polar regions does not melt; the air temperature is too low. Instead, the snow is stored, changing to glacial ice. As this process continues through time, the ice masses at the poles constantly grow in volume.

Introduction

As the prehistoric ice caps grew larger, they tended to throw the rotating planet off balance because of the wobble of the earth, causing the earth to roll around sideways to its direction of rotation.

Another analogy will make this clear. When you place a weight at the end of a string and then rotate the string in a circle, the weighted end of the string rises to a horizontal plane. Now, imagine yourself and the string as the earth, the weight at the end of the string as the weight of a growing ice cap, and imagine that, instead of intentionally swinging the weighted string, the rotational motion encompasses you, the string and the weight, as though you were standing on a rotating platform. In this depiction, then, your body represents both the present Axis of Spin and Axis of Figure of the earth. Your body does not move; the Axis of Spin remains the same. But your arm and the weighted string, here representing a radius of the earth, rise from the vertical (directed towards the pole) to the horizontal (directed towards the Equator). The sphere of which your arm and the weighted string are a radius is rolled sideways; the weight, representing the imbalance of an ice cap, rotates from a polar position to an equatorial position. The Axis of Figure, previously represented by your vertical arm, is now changed; the old Axis of Figure is now perpendicular to the Axis of Spin.

The rotating equilibrium thrown off balance by the weight of the growing ice caps, causes the spinning globe to roll over on its side. But such an event does not occur lightly. The oceans, like water in a bowl that is suddenly moved, are cast from their basins to flood the land. The winds, previously settled into patterns dependent upon a stable globe, are whipped asunder by the sudden shifting of the globe. The sudden meeting of warm and cold air creates great pressure zones that spawn new rains and hurricanes to sweep across the earth. The forces of nature, loosed from their equilibrium, rage wildly in search of a new equilibrium.

The Great Flood of Noah's day resulted from the latest careening of the globe. As we shall see later, records indicate that he had been living in what is now Madagascar or South-

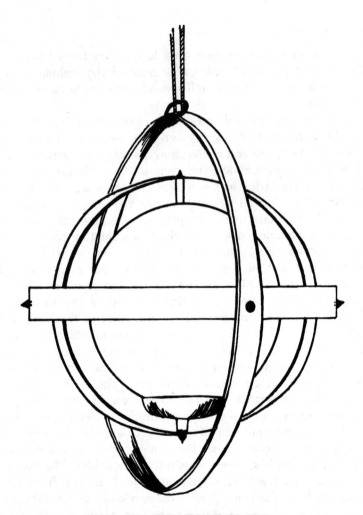

THE DOOMSDAY TOY GLOBE

Fig. 2. A small wooden globe representing the earth, with a small lead weight at one pivot representing an oversize polar ice cap at the South Pole, is supported in trunnions so that the globe may move in any direction. When a horizontal spinning motion is imparted to this laboratory model by the untwisting of the suspending cord, the lead weight promptly moves to a horizontal position, demonstrating the careening motion of the earth.

Introduction

east Africa—then about as far from the North Pole of Noah's epoch as New York City is from the North Pole of our day. At that time a land area containing the ice cap was at the North Pole. And on that day the globe careened, without changing its speed of rotation, through about 76 degrees of latitude. The ice cap rolled to what is now called the Sudan Basin of Africa, where it simply melted in the sunshine of the tropics, leaving tracks of its flow-off.

Every continent contains many groove marks of prehistoric ice ages. The slithering movements of towering glaciers have scoured the rocks over which they flowed, leaving a permanent record from which we can reconstruct their travels. And the groovings, all radiating from the center of the glacial areas, dis-

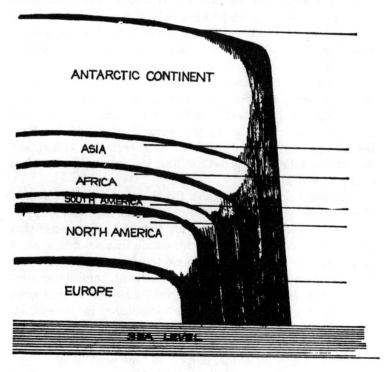

Fig. 3. Heights of continents.

close the location of the former North and South Pole areas, corresponding in size to the areas now contained within the Arctic and Antarctic Circles.

The present ice cap in Antarctica is merely the last of many thousands that have previously existed. Geological records reveal that it is the successor to a long lineage of glistening assassins of former civilizations of this earth. A minor ice cap also exists today, on the fringe of the Arctic Ocean, mainly in Greenland.

The enormous size of the South Pole Ice Cap is difficult to grasp. Were it centered in United States with the South Pole in North Dakota, its area would extend to the Atlantic and Pacific Oceans, to Mexico on the South and to the northern extremity of Canada on the North. The ice would stand two miles high at North Dakota, and icebergs would flow off into both oceans on a slope of about seven feet per mile. It would take one thousand four hundred cakes of ice the size of Lake Superior, with an area of almost 32,000 square miles, to equal the ice mass now accumulated on the Antarctic continent!

As large as the present ice cap is, however, there are two factors presently functioning to prevent, or at least inhibit, any immediate cataclysm similar to the ones of earlier epochs. The stress however, must be on the word "immediate," for these factors, one favoring the stability of the earth in rotation, and the other limiting the growth of the ice cap itself, existed in earlier epochs as well.

The earth only approximates the shape of a perfect sphere. The diameter of the earth at the poles is shorter than the Equator, a fact which suggests a slightly flattened sphere somewhat the shape of an apple. It is for this reason that when land areas on the short or polar axis roll around to where the long axis had been, as happened during earlier cataclysms, some land areas go below sea level. A new arrangement of land and ocean areas and a new equilibrium or isostasy are established for the ensuing epoch. By the time the careening and the flood are over, the top strata of the earth become readjusted to the new Axis of Figure and a new equatorial bulge is in time established.

Our globe today is stabilized and held to its present Axis of

Introduction

Figure by the centrifugal force of this equatorial bulge. The gyroscopic energy of the rotating bulge steadies the globe and prevents it from rolling haphazardly. The earth thus functions

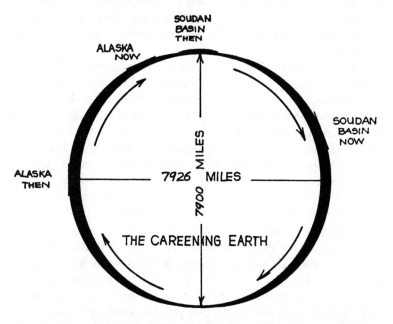

FIG. 4. Cross section (greatly exaggerated) of the bulge of the earth during the previous epoch of time. Alaska was on the 13-mile bulge of the earth then. It is now roughly 10 miles nearer to the earth's center.

The Sudan Basin Ice Cap was on the shortest diameter of the earth, at the North Pole. The Sudan Basin is now near the longest diameter of the earth, near the equator, roughly 10 miles farther from the earth's center.

in the manner of a flywheel. This mechanical energy acts like one jaw of a nutcracker whose other jaw is the energy of the "throw" of the eccentric centrifugal force of the rotating polar masses. The energy of these two jaws pinches certain of the upper layers of the earth strata until the pressure exceeds the resistance offered by the materials of the earth and causes a crumpling and elevating of earth layers in some places. The energy of the "throw" increases with the square of the speed of motion which in turn increases directly with the distance off center, thus overcoming the stabilizing effect of the earth's bulge and causing the globe to roll sideways to its direction of rotation.

Although this bulge is important dynamically to the maintenance of the equilibrium of the globe and the formation of the earth strata, it is smaller in size than is generally realized. Limited by the force of gravity, equal and opposite in direction, the bulge has a maximum of about 6⅔ miles at any point along the Equator, with an equal shortening at each pole. Since the best of ivory billiard balls warp from their tendency to absorb moisture, the earth is actually a truer sphere than a billiard ball, being true within about one-sixth of one per cent.

The other factor preventing an immediate cataclysm is a function of the dynamics of the ice cap itself. At present the glacial ice of Antarctica flows by its own weight and pressure, streaming through valley openings in the coastal mountains to the oceans where it flows away as icebergs. This disintegration, plus normal evaporation, provides a safety valve which gives our generation a chance to live on this earth. But only temporarily.

The discovery that the great South Pole Ice Cap is growing, rather than waning, as previously supposed, confronts us with an entirely new understanding of the limited time during which our present civilization has been developing, and the precariousness of its continuation. We are faced with the alternative of limiting the growth of the ice cap or accepting a limit to the duration of our present epoch.

The growing South Pole Ice Cap has become a stealthy, silent and relentless force of Nature—a result of the energy created by its eccentric rotation. The ice cap is the creeping peril, the deadly menace, and the divinely ordained executioner of our civilization. Just as a sword, suspended by a single hair, hung above the head of Damocles at Dionysius's banquet, so

Introduction

today the baneful jeopardy of an impeding world flood hangs over all of us.

This book will make it clear that, if we wish to continue to inhabit the earth we must control the further growth of the great South Pole Ice Cap.

The elemental forces of Nature that are involved are now known and they are clearly identified. If we procrastinate and do nothing, the Flood will occur when the present polar areas move away from the earth's Axis of Spin, and the Poles of Figure move to latitudes of ten to fifteen degrees, or about 5,500 miles away from the North and South Poles of Spin. The earth will tip over, like an overloaded canoe towed in a circle behind a power boat, in consequence of the wobble of the earth and the resulting eccentric centrifugal force of rotation of the present South Pole Ice Cap and its constantly increasing weight. The earth of today may quite readily be compared to a top-heavy, dying out, wobbling, spinning top, getting ready to fall over on its side.

PART ONE

I

Evidence of Careenings of the Globe

Historical Writings

FROM early Jewish history, as recorded in the Bible, comes down to us a tale of a great deluge, a tale familiar to most civilized peoples.* All mountains were covered by water. Noah and his family—together with two of every species of bird, beast, and reptile—were saved in an Ark which landed on Mt. Ararat in Asia Minor. The highest elevation of this mountain is 17,100 feet above sea level.

The physical cause of the Great Flood is confirmed by the biblical story, which consists of two merged narratives; from these we learn that "the same day were all the fountains of the great deep broken up. . . . And the rain was upon the earth forty days and forty nights."

It is both logical and evident that the breaking up of "all the fountains of the great deep" was the effect of a specific cause, and that the cause was a mechanical one, forcing a change in land and sea levels.

---

* Obviously there are still many people who consider the Noah tale as mere fiction. However, there are just as many who credit it with a certain historical truth; this writer is one of that group.

It is obvious that the rain of forty days and forty nights was incidental to the Flood, and not its cause. Noah and his group of survivors were "shut in" the Ark, and therefore knew nothing of outside atmospheric conditions. They may have thought that the rain caused the Flood, for the story of the rain has been passed down by their posterity.

Simple arithmetic shows the impracticality of the theory holding that the earth was flooded to the tops of all the mountains by rain water; this leaves "the fountains of the great deep" as the natural cause of the Flood, this being in accord with the historical record.

Mount Everest is 29,000 feet above sea level, and its top was submerged. In forty days and forty nights there are 960 hours, or 57,600 minutes, for the waters to rise 29,000 feet; thus the waters rose 725 feet per day, 30 feet per hour, or approximately 6 inches per minute!

Obviously no continuous rainfall could create so great a flood. Since rain waters would run off into the oceans such a flood would be impossible by means of rainfall. Further, it would be beyond the capacity of the rain cycle of evaporation, condensation, and precipitation to produce it.

Therefore, the fountainous breaking up of the waters of the great deep,* caused by the movements of earth materials, remains as the logical interpretation of the biblical story of the great Flood. Furthermore, this clearly fits into the pattern of the careening globe theory, and aids in identifying that last careen of the earth as the cause of the Flood.

In the mythology of the Greeks, the iniquity of the human race provoked Zeus to overwhelm the earth with a flood; this occurred, it was said, in the fifteenth century before their era. From this flood, only one man, Deucalion, and his wife, Pyrra, survived in an ark or chest which came to rest on Mt. Parnassus, Greece. The elevation above present sea level of this mountain is 8,000 feet. The Hellenes (Greeks) were descended from Deucalion's son, Hellen.

---

* See pages 51, 144.

Evidence of Careenings of the Globe

The ancient Hindus, Chaldeans, and the Jews all have records indicating that a great deluge occurred slightly more than 5,000 years ago.

Cuvier refers, without identification, to an ancient Brahman collection of Indo-European prose which had a recurrent flood theory.

William Thomasson says in his book *The Glacial Period and The Deluge* that "the Chinese have a wonderful tradition, that properly interpretated, tells of their sudden, flying leap to the Arctic. . . . It is the story of the Ten Stems, or Ages."

Confucius, born about 551 B.C., begins his history of China with a reference to a receding flood which had been "raised to the heavens."

Plato relates through Critias the story, told to Solon by Egyptian priests in 600 B.C., that a great war of invasion had occurred about 9,000 years earlier, led by the kings of Atlantis, an island empire of very great extent, which was afterwards sunk by an earthquake and left an impassable barrier of mud to sailors voyaging past what is now Gibraltar.

ATLANTIS—*The Antediluvian World*, by Ignatius Donnelly, contains historical proofs of a great deluge, including details of many written records and legends of Assyrian, Babylonian, Chaldean, Hindu, and North and South American origin.

Mr. Donnelly refers only to the disappearance of Atlantis, and with it, its civilization. His researches disclose evidence of a civilization prior to the Flood, and of the dispersion of peoples and their arts following the catastrophe. Actually, most of the then existing peoples and their civilization were engulfed immediately by this latest World Flood.

Theories about the earth's so-called crust of one hundred to one hundred fifty years ago reveal that the science writers of that period appear to have been divided into two main groups. Those who belonged to the older school of thought were referred to as "Cataclysmists," or "Catastrophists." They held to the inherited, then accepted, theory that the main changes observable in the earth's surface were the results of an adjusting power different from what is now commonly understood as The Laws of

Nature. In the view of the newer school of thought, represented by those referred to as "Uniformitarians," the workings of natural, unchanging, explainable forces could account for all of the changes in the surface of the globe. Since then, the scientific world has become, in its beliefs, wholly uniformitarian; the cataclysmic theory of geological changes through the caprice of nature has been gradually abandoned. The missing link, which kept cataclysmists and uniformitarians separated, is simply an acceptance as a basic scientific truth that catastrophism is a part of the natural working of Nature's Laws.

A hundred odd years ago cataclysms were recognized as having occurred—as they are so recognized today—but the explanations offered were not generally accepted because they were based on the so-called caprice of nature: erratic, freakish, whimsical control by Nature—or the temperamental setting aside of Nature's Laws. Today, scientists look to the Laws of Nature for explanations of all physical phenomena.

The difference between cataclysmists, who claimed to know the answers, even though wrong, and the fundamentalists, who were still seeking for a scientific explanation to which they might agree, is illustrated by the following story.

A cataclysmist was asked how he would explain the phenomenon if he saw a bar of steel floating in the air. "Why," he said, "if I happened to witness such a thing I would know that it proved the temporary suspension of one of Nature's Laws."

A fundamentalist, when asked the same question, replied, "If I saw steel floating in the air I would know it proved the existence of a Law of Nature about which I happened to be ignorant."

What the cataclysmists explained erroneously, at the time, and the uniformitarians left unexplained, is now rationally explained by the basic theory of an automatically careening globe—a theory which is strictly uniformitarian, being wholly in accord with the immutable Laws of Nature.

Archeology

AT THE site of the ancient city of Ur of the Chaldeans, located in present-day Iraq about eight miles west of the Euphrates River and near its junction with the Tigris, archeologists have disclosed layers of materials which indicate that one city after another occupied the area during a long period of time. Excavating layer after layer to a depth of about fifty feet, they have disclosed about one hundred and thirty-five successive periods of city life, each period also being demarcated by a different dynasty.

At slightly below the depth of fifty feet the archeologists came upon a layer of clay, eight to ten feet thick. Below the clay bed they discovered ten (one account says twelve) layers representing successive dynasties, but the relics and artifacts were found to be of a different kind than those found above the clay. A significant discovery was the fact that painted pottery was found below the clay bed but not above it, with the exception of scattered samples of painted pottery found immediately above the clay, but not higher up.

The archeologists are in agreement that a flood must have produced the bed of clay. Clay is formed by silt settling in water. The silt is derived primarily from the grinding of rocks upon rocks under the pressures and movements of glacial ice.

A ten-foot clay bed took a long time to develop; the length of time required for its creation may be ascertained by counting its varves or layers—as explained in the section entitled "Geology."

Copper is an ingredient of the articles found in the strata above the clay bed, but copper is absent from the artifacts below—suggesting a discontinuity of the two civilizations separated by the period of time during which the clay bed was being created.

By comparing the relics and artifacts found in the different layers of dirt with those of other civilizations, we can estimate the elapsed time represented by the total fill of dirt above the clay bed to be about 6,000 years.

A high degree of civilization, at the culmination of Epoch No. 2 B.P., (Before Present) would account for the painted pottery. A movement of the land area, caused by the careening of the earth, from a temperate to a cold latitude, its submergence beneath the seas and the entire disappearance of its people and their civilization, is indicated for Epoch No. 1 B.P., during which the clay bed was formed.

At the commencement of the present epoch, at the moment of the last great flood, the land area careened to its present latitude, where it again became inhabited by human beings; but these new peoples did not possess the art of painting pottery, which characterized the artifacts of the race that had been destroyed by the flood.

Ur of the Chaldeans was probably located on the Persian Gulf, though it is now about 115 miles inland. The sea level has not remained constant, as explained elsewhere, and the lower delta of the Euphrates River has extended into the Gulf. Clues to this are the record found at Ur of a marine hero, conqueror of storm and sea, and artifacts indicating trade with distant places—probably partly by sea.

Excavations by archeologists at Cnossus, Crete, have disclosed 43 feet of soil and then virgin rock. The relics and artifacts uncovered indicate successive habitations by man, and when compared with specimens gathered elsewhere, these objects represent a time period of about 5,300 years.

The soil which developed during these distant years came from vegetation and animal remains, from wind-borne dust, and from erosions at higher levels. The relics and artifacts were not found in the very deepest excavations near the rock surface; this fact indicates that a period of time elapsed before people came to live there and also that the gradual build-up of the earth materials had extended over a period of more than 5,300 years.

The rock substrata below the soil at Cnossus fits into the pattern of the theory of a careening globe. A revolutionary change in the development of the earth strata occurred at this place.

During Epoch No. 1 B.P. the land area now known as the

island of Crete was located near a latitude corresponding to the present Arctic Circle. Today, located near the eastern end of the Mediterranean Sea, it is in a temperate climate, is covered with soil and vegetation, and people have been living there for more than 5,300 years.

The Mammoths

ANIMAL fossils, especially mammoths, offer positive proof that the earth has rolled around sideways to its normal direction of rotation.

Mammoths are now being found in arctic regions, buried in lifelike condition in the permanently frozen ground. Their presence, condition, and location document a gigantic catastrophe in which the climate of a very large area of land suddenly and drastically changed. Only a sudden rotating or careening of the globe could have caused this change.

Siberian mammoths are commonly described by the terms "Wooly" or "Northern." Their bodies and limbs are covered by long coarse hair resembling tubular reddish-brown plastic needles. In addition, they have been found with an undercoat of short finer hair. But the skin of their head, trunk, and ears is smooth. It is evident that they, like our present-day elephants, were unsuited to a cold climate. Indeed, they would have frozen solid in the present winter climate of Siberia.

The food contents of their stomach and their teeth, however, provides us with evidence of their origin.

The literature on the mammoths is full of references to their eating evergreens, now the main vegetation of the regions where the carcasses are found. This reference has apparently resulted from the writers' assuming that the mammoths actually lived and thrived in the present cold climate of Siberia. Analysis of the stomach contents of these carcasses does not substantiate this theory. The evidence suggests that the feeding grounds containing these animals was moved quickly from a warm to a frigid climate. The carcasses of rhinoceroses, also found in the ground, aid in further substantiating that some of the feeding grounds were tropical, and that the present polar climate of their resting places is far different from the climate of the Eden-like land in which they were born and reared.

The food contents of the stomachs of the mammoths also give us a clue as to the exact time of day the earth careened. A full stomach indicates the sudden death of a healthy animal, and that death occurred after, and not before, the feeding period of the day. The food found in the mouths and stomachs of prehistoric monsters indicates that they had been grazing among abundant warm climate grasses when death suddenly overtook them.

The suggestion of a sudden careening of the globe is further substantiated by the condition of the carcasses. Several mammoths have been found in an upright position on their haunches. Some have been found with broken bones. The upright position supports the theory that they met death suddenly; the broken bones indicate violent contusions just prior to death. The superhurricanes, or head winds caused by the rapid careening of the globe, would account for large animals being tossed about and buried in debris. The raging waters of a flood would produce a similar effect. The lifelike condition of mammoths found underground would indicate that they were frozen solid soon after having been buried alive.

In 1901, a mammoth was extricated from the bank of the Bereskovka River in Siberia, 66° N. latitude, almost on the Arctic Circle. It was solidly frozen in the tundra, but its head became exposed during a landslide. It was a male animal, found sitting on his haunches, with pelvis bone and right foreleg broken. In this condition he could not move, much less forage for food. Yet it had perished just after eating breakfast. There was a small quantity of grass on his tongue which he had been in the act of eating. His teeth were filled with half-chewed grasses; twenty-seven pounds of grass were removed from his stomach on one occasion, and more on another. This animal is now mounted in the Zoological Institute of The Academy of Sciences in Leningrad.

The stomach contents of the Bereskovka mammoth consisted

Evidence of Careenings of the Globe

chiefly of field grasses, which were identified, analysed, and photographed. The names of the grasses are given in Russian and Latin in an article by G. N. Kutomanov in the *Bulletin of The Academy of Sciences of St. Petersburg,* 1914, Vol. 8, No. 6, pages 377–88. The grasses are similarly reported in a detailed description of the extrication of the beast in the annual report of The Academy of Sciences, 1914, Tome 13. That report observes that "Contrary to popular belief, no evergreens have ever been found in the stomach of a mammoth."

Nine genera of grasses were found and help us to establish the climatic conditions under which the animal lived. If the grasses were arctic grasses, the mammoth must have lived in an arctic climate. If the grasses were tropical, a tropical climate would be indicated. This problem was submitted to the Smithsonian Institute. Mr. C. V. Morton, Curator, Division of Ferns, Department of Botany, advises that all of the grasses are now found in temperate climates, none in tropical climates, and four out of the nine are found as far north as the Arctic Circle.

Whether the grasses could have grown in a tropical climate, and survived after having been moved to temperate and frigid climates, is not ascertainable. The presence of rhinoceroses, however, indicates that the climate had been tropical.

A report concerning a rhinoceros found on the bank of the Vilui River in Siberia states: "The animal appears to have been drowned, for the blood vessels of the head were found by Professor Brandt to be filled with red coagulated blood, such as would be produced by suffocating through drowning. Probably it was suddenly caught in a flood of rushing water, from which it had no opportunity to escape. At one moment the animal was standing on firm ground, peacefully browsing, and in the next was overwhelmed by a roaring flood, the tumultous waves of which bore along masses of mud and gravel in their sweeping course, so that it was drowned and buried almost instantly. Then the intense cold set in, the body froze, and the ground never thawed out until the day when it fell down on the banks of the river."

Both the Vilui River rhinoceros and the Bereskovka River

mammoth evidently died of suffocation. The stomach contents of the mammoth, as indicated by the many photographs, did not contain water. The grasses were dry. Therefore, it is reasoned, the beast was not drowned but perished in the super-hurricane and dust and dirt storm caused by the rapid movement of the earth's surface against the air in that particular area. The same winds, by their force and pressure, would have filled the air with the trees, animals, top soil, sand, gravel and debris, in which the animals were buried alive.

I. P. Tolmachoff states concerning the Bereskovka mammoth: "The pelvis, a right foreleg and a few ribs were found broken, as well as indications of a strong haemorrhage and also suffocation in mud. The death by suffocation is proved by the erection of the male genital, a condition inexplicable in any other way." (*American Philosophical Society Transactions, N.S. 23, 1929*).

Physicians have corroborated Mr. Tolmachoff's conclusion of suffocation; this conclusion, in turn, helps to establish the fact that these animals died through sudden mass extinction, and not by slow or individually separate deaths. Tolmachoff also states that no mammoth nor rhinoceros has been found frozen in the ice.

The fossil remains of other beasts and fishes have been found with undigested stomach contents. A beast with a partly chewed rodent, for example, was found in Colombia, South America, in 1945, and is now at the University of California, in Berkeley. This beast, classed as genus Borhyena has been estimated to be millions of years old. It had been buried in fine sand before it had had a chance to digest its recently swallowed breakfast. When the sandstone was carved away from the skeleton, the rodent was found resting where the beast's intestines belonged.

The arctic regions, where mammoths, rhinoceroses, and other animals have been found, do not have sufficient vegetation to support a single mammoth, and the cold is so intense in winter that no mammoth could survive. Yet, just prior to the latest careening of the globe this region was populated with teeming herds of animals.

They lived there because an ample food supply existed, and

Evidence of Careenings of the Globe

the food supply grew because the climate was warm. Millions of mammoths once lived in what is now a refrigerator for their carcasses and bones.

The abundant vegetation, indicated by the food supply, corroborates the other evidence that the latitude where these animals lived was either tropical or temperate.

Great quantities of bones of mammoths, horses, cattle, buffaloes, camels, sheep, deer, and many other grass-eating animals—as well as those that preyed on the plant eaters—have been found in the frozen tundra of Siberia. Their remains add to the positive evidence of the profuse vegetation necessary to support these hordes of animals. The finding of at least thirty-nine mammoths in the Siberian tundra is recorded.

Animals smaller and less spectacular than the mammoths and rhinoceroses have not been reported, when and if they have been found by hunters and trappers. Nevertheless, a great number of smaller animals must have become exposed on the surface through tundra landslides caused by summer rains that, unable to penetrate the frozen tundras, flood extensive land areas.

In regard to the remains of mammoths, mastodons, dinosaurs, and other prehistoric animals now being found at widely scattered areas of the earth, at many different latitudes, and in successive earth formations, three facts stand out: First, the fact of their total destruction. Second, the fact that the last members of the species died suddenly while in a condition of good health. Third, the fact that their remains show their habitations extended over widely scattered and now separated land areas.*

The theory of the recurrent careenings of the globe fits the evidence better than any other. The careening theory explains the cataclysms destroying animal and plant life, and accounts

---

* The moot question of a land bridge at Bering Strait, between North America and Asia, is apparently solved by the great quantities of mammoth tusks and bones found in the now separated and frigid areas of Wrangell Island, New Siberian Islands, Alaska and Siberia; indicating that these animals roamed freely over a connected land area, in a warm climate, just previous to the latest careen of the globe.

for changes in the climate of most areas of the earth as well as the duration of each epoch between the world deluges.

Sea Life

THE seas have also been searched for organic life which would help substantiate the theory of a careening globe, and clues have been found in seals and lobsters.

The seals found in the Caspian Sea and in Lake Baikal in Siberia are the same as the seals which inhabit Alaskan waters. The evidence indicates that the two branches of the family at one time were together, like the mammoths, and became separated during the last great deluge. Most of the lakes—as well as land areas—of the globe were then temporarily covered with the waters of the oceans enabling the seals during the Flood to scatter in all directions.

There is a logical, self-evident explanation to the riddle why the same variety of seal happens to be found in three such widely-separated locations. They are the descendants of those seal ancestors that were still living, and could find a food supply, when the Flood came to an end. Some among the ancestor group of seals had been stranded on land, some found themselves in lakes, while others were still in the ocean.

A lobster of peculiar genus is found only in icy arctic waters and in the Pola Deep of the Mediterranean Sea. Finding this lobster in the Mediterranean Sea helps to prove that the sea was near the North Pole before the last careening of the earth. At that time its waters were icy and suited to this species of cold-water lobster. When the earth last careened this sea was moved to a temperate climate. The cold-water lobster still continues to live in it, but only in its coldest waters and in its deepest depression.

Fossils

ABOUT a century and a half ago Georges Cuvier wrote: "It is to fossils that we owe the discovery of the true theory of the

Evidence of Careenings of the Globe

earth; without them, we should not have dreamed, perhaps, that the globe was formed at successive epochs, and by a series of different operations. They alone, in short, tell us with certainty that the globe has not always had the same envelope; we cannot resist the conviction that they must have lived on the surface of the earth before being buried in its depths; if we had only unfossiliferous rocks to examine, no one could maintain that the earth was not formed all at once."

There is, today, among scientists complete agreement with Cuvier. Drillings to a depth of four miles have disclosed the earth's envelopes, now called strata, and each provides us with a record of the epoch during which it was created.

Footprints and tracks of animals, reptiles, and crustacea, that were made many thousands of years ago in various muds and wet sands, have been discovered and are now preserved in museums in the form of rock specimens. Raindrop splashes in the then soft, oozy mud have been discovered in numerous specimens of stone.

Where the evidence of a tropical climate surrounds frozen mud sculptures the sudden freezing can only be accounted for by an assumed careening of the earth which brought the mud into a different climate. The prompt solidification of the mud by freezing, when moved quickly from a tropical or temperate climate into a frigid climate, clearly accounts for these remarkable phenomena.

The mud sculptures—having become like stone by freezing—were further "set," during one of the glacial periods, by the accumulation of a layer of sediment. This, in turn, acted as a mold and preserved the shapes of the sculptures after the specimens had been careened back to tropical or temperate climates and the frozen mud or tundra thawed out during the succeeding epoch. In these molds the former mud slowly changed to stone.

There are great differences in the fossil markings on rocks. The sharp, delicate, shell-like craters of raindrop splashes could not have been preserved except by quick freeze; the mud of unfrozen splashes soon oozes back and become pockmarks. Jellyfish entombed in mud and frogs could not have been preserved except by quick freeze; else they soon would have rotted.

On the shores of the Bay of Fundy large areas of dried red and sandy mud, deposited by spring tides, are laid bare and, baking in the hot summer sun for ten days during neap tides, the upper part of the mud becomes consolidated for a depth of several inches. Sir Charles Lyell reports finding, on these mud surfaces, small cavities or pit marks caused by raindrops, footmarks where birds had walked, and other tracks. On splitting a sample slab of the hardened mud and reducing it in thickness, he found footprints made during several prior neap tides on the inferior layers: each made by birds at different times.

This is an illustration of one of Nature's methods of preserving trackmarks. Even today fossil prints of various kinds are produced in this way and may be found in many different localities. This drying-out method of producing stone from mud, and preserving trackmarks, could never account for the delicate shell-like craters of raindrop splashes which are preserved by quick freeze.

The most sharply delineated markings of tracks of living things and imprints of vegetation preserved in stone are found in the top layers of the strata that correspond to the end of each epoch. For example: the profusion of leaf and fern details on the top surfaces of vegetable muck deposits which have become coal (as described in detail later), indicates solidification by freezing during the last moments of the thousands of years of muck accumulation; after that moment no more muck accumulated at that location.

Where these muck deposits were located, a revolutionary change occurred in the way the earth's materials are formed. From this we know that the conditions necessary for the forming of those earth layers suddenly changed. We know that a condition for the formation of vegetable muck was a tropical or temperate climate, and we know that its slow, time consuming creation suddenly ceased. We know that it would disintegrate or be consumed by slow combustion unless it was suddenly covered up.

Evidence of Careenings of the Globe 27

We therefore look for evidences of a polar climate in the overlying strata, and we often find clays, shales, and slates, which confirm the theory of a careening globe.

A confirmation of the theory of a careening globe—evidenced by the time elapsed between successive layers of the earth's upper strata—comes from the tracks of large dinosaurs which were examined by a trained observer, in 1940, on the Davenport Ranch in Bandera County, Texas. He reported sun cracks in the silt filling the footprints; this indicated that the surface had been below, although it is now above, water.

The important element in that observation is the fact that the sun cracks were in the silt filling the imprints, but not in the rock materials containing the track marks. From this it may be readily deducted that the tracks had become set as hardpan and then rock before the silt filler that cracked in the sun had been deposited; and that, therefore, the silt filler represented a later epoch of time than did the rock material which did not crack in the sun, and could not have been contemporaneous.

The dinosaurs whose skeletons were found grouped together in the rock formations at Dinosaur Monument, Utah, were drowned by the Great Deluge which ended the epoch in which they lived; they sank to the bottom of a lake or river and became covered with sediment which turned to rock during succeeding epochs of time.

Those rocks are now small mountains. The upheaval raising lake or river bottoms to much higher elevations occurred during one of the later careenings of the earth with simultaneous rearrangements of land masses and a Great Deluge. From Dinosaur Monument a million pounds of petrified bones have been quarried for display in various museums.

Mass graveyards with remains of mammoths have also been discovered. Geoffrey Bibby, in his book *The Testimony of the Spade*, describes one graveyard containing over 900 mammoths, both young and old, several hundred other grazing animals, and wolves and foxes; it is located at Predmosti in central Moravia, in a valley quarry six to ten feet below the surface of the covering dirt and top soil, in a stratum which has not yet turned to rock. He states some conflicting opinions as to the probable reason for its existence. He cites similar mass burial grounds of mammoths as having been found in Lower Austria, at Krems, Langmannerdorf, and Willendorf, and elsewhere. Evzen and Jiri Neustupny, in their book *Czechoslovakia Before the Slavs* (page 26) state that "the bones of more than a thousand mammoths have been found at Predmosti and the quantities discovered at Dolni Vestonice and Pavlov are no less impressive."

The mammoths' graveyards can be considered as additional evidence of the recurrent cataclysms of the earth. Their shallow burials make it appear probable that they lived in Epoch No. 2 B.P., when the Hudson Bay Basin was at the North Pole of Spin; it also seems probable that their carcasses have not received quite as much protection against disintegration as have the New York State mastodons of Epoch No. 3 B.P.

An exhibit at the American Museum of Natural History showing a similar group of skeletons of prehistoric animals, all piled together like offal at a slaughterhouse, can be explained most rationally by the deluge caused by a careen of the globe. Those animals evidently came to their death by cataclysmic mass drownings. Their bodies probably settled in an eddy, or at an obstruction, or in a deep hole at the bottom of the transient flood waters, where they were covered by dirt and debris. Quick freezing may also have retarded their disintegration.

The careening of the globe, with concurrent great deluges, is confirmed by such discoveries of massed skeletons of contemporaneous animals piled together. Similar burial grounds containing contemporaneous fish skeletons will be discussed later.

Petrified oysters, clams, crabs, and starfish were found at depths of several hundred feet during the digging of the Panama Canal. They were all perfectly preserved but had turned to stone. Some of the species do not thrive in the tropics, indicating that what is now Panama was at one time located in a temperate zone.

Specimens of fossil jellyfish have been discovered in Cambrian rock formations, classified as among the oldest rocks. Their external structures, as well as something of the interior forms

Evidence of Careenings of the Globe

of the jellyfish, were found to be quite well preserved. (*Geology*, by H. F. Cleland, page 416.)

Solidification by freezing of both the sand and the jellyfish, at the moment of the careen of the globe, is the simplest scientific answer to this age-old riddle of how a soft jellyfish could become solid rock. What is now rock was once soft and wet sand which was suddenly hurled about so that the jellyfish was virtually buried in it; thereupon both suddenly congealed into a solid mass by quick freezing.

The preponderance of marine fossils found so far, as compared to upland fossils, is partly due to the cleavages of the unconformable debris which covers former sea surfaces. Such cleavages bring about the exposure of the trapped and preserved specimens, among which the best preserved are those that have been quickly frozen. The fossils of the uplands embedded in what was soil at that time, are less easily discoverable. Many animals, including dinosaurs, are found in rock. Mammoths are being found in tundra or dirt that will change to rock, and mastodons have been retrieved from moist earth, well below the surface, which will eventually become rock.

As we have seen, the bones of fossil animals must be assumed to be those of animals quickly buried after death, for bones left on the surface decay rather rapidly due to oxidation and the action of organic acids wherever vegetation flourishes. The former enormous herds of buffalo on the American plains did not become fossils. The present swarming animal life of the African plains does not become fossilized at death. At death animals become part of the substances building up the soil. But when mass burials of animals have occurred—due to the careenings of the globe—the remains have been embedded in earth through which mineral-laden waters have percolated and have established conditions for creating fossils.

Trees and Vegetation

FOSSIL trees and other vegetation provide additional evidence regarding past epochs of our planet. Upright trees and tree trunks are found in the sea; fresh tree trunks lie underground; fresh fruit and leaves, frozen like the mammals, are found in Siberia; fossilized and petrified trees exist not only on the surface of the earth, but also in tiered layers of its sub-strata. And all of these phenomena can be traced quite rationally to the past careenings of the globe.

In the Bay of Fundy, at Fort Lawrence, Nova Scotia, the stumps of a submerged forest of pine and beech trees stand upright in the soil in which it once grew. They become visible during low tide. In other parts of the Bay of Fundy also, short, decaying stumps and roots emerge briefly and are exposed to view during low tide.

These trunks and stumps are the remnants of trees once growing in upland areas that were completely submerged when the earth last careened. As the sea level later was lowered (due to the waters accumulating as ice in Antarctica and elsewhere) the trees were all gradually exposed to the air. Oxidation occurred, and the exposed parts rotted away.

The tree trunks of any submerged forest all end abruptly at low tide water level, for any part of a tree projecting above the water or mud would, if given enough time, become oxidized by the air, would rot, and be washed away. What will finally be left are the stumps and short trunks standing below the lowest water level.

Tree branches in a vertical position were reported by Nordenskiöld (in *The Voyage of Vega*) to be at the bottom of the sea adjacent to the arctic islands of New Siberia. Nordenskiöld also refers to tree branches which burn with a glow, without a flame, and which continue to be cast up every year in a northern Siberian lake, indicating submerged forests, beneath the surface of the lake.

At many places tree trunks have been found underground. These trees obviously did not grow underground, and under normal conditions no fallen tree becomes buried. They must have grown above ground in some former epoch and then been buried by a cataclysm, for dead trees lying on the ground merely rot and decay. They are gradually oxidized, just like the decay-

Evidence of Careenings of the Globe

ing tree limbs projecting above water. Under such conditions they would have disappeared entirely before a hundred years had elapsed. But when trees are buried in water or damp earth they are protected from oxidation, and are able to stay fresh for thousands of years. The presence of these underground trees is further evidence of a cataclysm that buried them under dirt and debris borne by hurricanes and flood waters.

Fresh trees can now be mined in many places, including the Dismal Swamps of Virginia, the Hackensack, New Jersey, meadows, and in the marsh area of the isthmus connecting Nova Scotia and New Brunswick.

In certain areas of northern Siberia innumerable tree trunks—called by the natives "Adam's wood" and said to be in all stages of decay—are embedded in the solidly frozen tundra. Because they were once growing trees, of types which do not grow in that climate, they confirm that a change in climate has taken place, such as would be caused by a careen of the globe. They could have been broken by a hurricane or flood. If so, they will show a clean break on the side on which the breaking force was imposed and torn fibers on the lee side. A reexamination of the wood, to determine genera and species of the trees, will enable us to establish the latitude range or climate in which these trees grew.

A so-called mammoth tree, with fruit and leaves still on it, was discovered and reported after a landslide of Siberian tundra. Such cold storage of fruit 7,000 years old can only be explained by a sudden transportation of the fruit from a warm climate in which it grew to the cold storage climate in which it has been refrigerated. This specimen of fruit, with leaves, and many other specimens of leaves reported found in Siberia also confirm the careen of the globe.

The American Museum of Natural History in New York possesses an exhibit of fruit and plant fossils postulated as millions of years old; the exhibit includes figs and palm fruits; fresh, full-sized banana leaves; fig, palm, sycamore, pine, and ginko leaves; sequoia pine cones, and water chestnuts. The fruits are full-size and luscious looking, as though freshly fallen from the trees; the leaves are also full and fresh looking, not shrunken or folded as from exposure to the sun, but appear as just fallen, or laid down in water.

It is necessary to apply the theory of "quick freeze" to these specimens, as otherwise they would have become rotted, crushed, or otherwise destroyed, like most other vegetation. The fruits and leaves "set" by "quick freezing," and then being hermetically sealed within soil which became rock, the prerequisites were established for the slow process of petrification to take place.

Nothing else accounts for the fossilization of this vegetation but the careening of the earth. Figs, for example, are a tropical or semitropical product. To be preserved they had to be frozen, and to become frozen they must have been moved to a frigid climate.

Fossil trees are found all over the world. Outstanding examples of petrified forests are near Cairo, Egypt, at sea level, and those high up in the Rocky Mountains in Yellowstone Park, near the continental divide.

At the latter location there are twenty-seven horizontal tiers of former tree life, representing an equal number of Life Ages; all have become exposed as the side of the mountain has been gouged out.

The fossil trees which have become exposed in some of the earth's layers show recurrent periods of tree life; the strata showing tree life are sometimes separated by strata of earth from which trees are missing. Where tree fossils occur in one stratum, are absent in the strata above and below, but occur in the next adjacent strata, they give us an authoritative confirmation of the careening of the earth. The slow rising and sinking of land areas relative to sea levels, which prevail at all times, cannot account for this phenomenon.

Upright fossil trees are found at many different levels at the Bay of Fundy, Nova Scotia. The tree trunks vary in diameter from fourteen inches to four feet, and in height from six to twenty feet. The lower ends are in strata of coal or shale. Tree

Evidence of Careenings of the Globe 33

roots penetrate two different strata in some locations. The tree trunks, all cut off abruptly at the tops, extend through different strata of shale, sandstone and clay, but never through a seam of coal above them. Tree roots having grown through two adjacent strata of earth confirm the assumption that they grew in soil and that the soil has changed into rock.

The coal and shale strata from which the trees sprouted are seen to belong to an earlier period than the superimposed strata above the upper ends of the vertical tree trunks. The seam of coal next above was a still later development. The beds containing the fossil trees are usually separated from each other by masses of shale and sandstone—many yards in thickness. These strata represent the developments of many thousands of years and successive epochs of time.

Nothing of the original trees is preserved except the bark; it forms tubes of pure bituminous coal and is filled with sand, clay, and other deposits which appear like solid internal cylinders. In one of the trees examined by Sir Charles Lyell nine distinct layers, or deposits, formed the interior cylinder, while there were only three layers of earth surrounding the tree. The formations in which the tiered layers of upright fossil tree trunks are found contain also about nineteen seams of coal. They range from two to three miles in length along the coast, and are not interrupted by faults.

The best view of these ancient tree trunks may be had at Joggins, where the cliffs are 150 to 200 feet high, forming the southeastern shore of an inlet of the Bay of Fundy, called Chignecto Bay. The fossil trees are all at right angles to the planes of stratification, which are inclined at an angle of 24 degrees to the south-southwest. The strike lines of these planes, together with the length of the uplifted formation, as shown in data from the Spur Ranch drilling (see page 73), indicate that the trees were once buried about 2 to 2½ miles below the surface, and that they are about five to ten million years old.

The circumstance that all the trees have been cut off abruptly at the tops suggests that they once stood under water as submerged forests, at which time the bark became carbonized—as is happening today to the upright tree stumps in the Bay of Fundy.

The tiered layers of fossil trees are a visual confirmation of the fact that the earth has careened repeatedly. Each layer was developed during a different epoch. Each epoch ended with a change of true latitude for the land area now known as Nova Scotia.

Similar tiered layers of fossil trees are found in the arctic regions. Frozen "Wood Hill"—in the New Siberia Islands, well within the Arctic Circle—is described by Nordenskiöld as being 200 feet high and consisting of thick horizontal sandstone beds alternating with strata of fissile bituminous tree stems, heaped on each other to the top of the hill, with vertical tree trunks embedded in the sandstone of the upper strata.

This scientific disclosure, on analysis, shows that the bituminous tree stems, which are now in the fossilized form of coal, were the successive growths of earlier ages. The thick sandstone beds correspond to sands created during the intermediate epochs of time, or sands left by one of the successive great deluges of the earth.

The fact that these fossil trees are displayed on the side of a hill indicates that the hill is a remnant of land left standing after the surrounding land areas were gouged out by flood or glacier. The vertical tree stems in the upper strata are the remnants of trees which were growing at the time of the latest cataclysm of the globe, and it can be predicted with confidence that, on reexamination, the protruding tree trunks will prove to be growths of a temperate or tropical climate.

Superimposed coal fields, separated by considerable thicknesses of rock, are described in "Fossil Flora of Sydney Coalfields, Nova Scotia," by W. A. Bell (in Memoir 215, Geological Survey of Canada). The text and illustrations describe hundreds of specimens of fossil leaves, ferns, tree bark, and wood. Many of the successive horizons contain duplicate fossils; but in each horizon differences in species occur, with the earliest (lowest) ones differing most from the latest.

Evidence of Careenings of the Globe

It is natural to assume that the fossil leaves and barks of trees came from trees that also have been fossilized. Layers of upright fossil trees, like those in the cliffs of Joggins on the opposite side of Nova Scotia, are never found in the coal seams, and it is therefore assumed that they would remain undiscovered in the drifts of the Sydney coal mines, to which Mr. Bell confines his report. His specimens were mostly taken from the roofs of coal mines which extend three miles out from shoreline under the sea. Fossil trees are exposed in superimposition on the adjacent cliffs, as at Joggins.

Similar vertical fossil tree trunks have been found in other locations. For example, at St. Helen's, Lancashire, England, they occur in silty clay below a layer of about seven feet of brownish-colored topsoil. The stratum containing the trees is reported to be about twenty-one feet thick, inclined twenty-three degrees to the east-southeast, and rests on white sandstone. The tree trunks begin on a level about 8½ feet above the white sandstone stratum and extend up about nine feet.

These trees grew during a previous epoch, but they did not grow in England. The topsoil, above the tree-bearing stratum, is all that was developed in the land area now known as England.

In all the cases discussed it is obvious that the trees grew to their present size upon the earth's surface, were suddenly buried in water, mud, or moist earth, and after fossilization, were returned to the surface of the earth when the covering strata of materials were gouged out by glaciers or washed away during great deluges. Again, the theory of the careening of the globe explains all the evidence we have here reviewed.

Rivers and Waterfalls

THE waterfalls of certain rivers furnish us with time scales with which we can estimate the duration of their existence.

In the language of geology, a waterfall, or cataract, is a temporary erosion in the land which is always moving upstream. This stems from the fact that the brink of the falls is being worn away continually by crumbling and erosion of the

Superimposed Strata of Stone Containing Fossil Trees and Fossil Flora

(In some cases definitely known to be separated by massive strata of non-fossiliferous rock)

Location	Number of tree-bearing strata	Type of fossilization	Reported by
1. Sydney Mines, Cape Breton, Nova Scotia	59	Fossil Flora	Geological Survey of Canada, Memoir 215
2. Yellowstone National Park	27	Petrified wood and bark	Longwell and Flint in *Outlines of Physical Geology*, 1962
3. Wales	17	Superimposed fossil trees	W. J. Fielding in *Shackles of the Supernatural*
4. Joggins, Nova Scotia	10 plus	Petrified bituminous tree bark surrounding stone cylinders	Sir Charles Lyell, in *Travels in North America in 1841–2*
5. New Siberian Islands, "Wood Hill"	Many	Fissile bituminous tree stems in strata alternating with thick sandstone beds; heaped on each other to the top of the hill	Nordenskiöld in *The Voyage of the Vega*

Evidence of Careenings of the Globe 37

earth materials of the ledges, and the lowest rock layers, below the falls, are being constantly cut away by the forces created by the falling waters. As a result of this erosion it is possible to estimate the length of the life of a waterfall and to determine the duration of our present epoch by the life span of the waterfall.

The falls of the Niagara River have moved upstream from what is now Lewiston, on Lake Ontario, and have created a gorge which is now about seven miles long. Records kept by the U.S. Geological Survey since 1842 regarding the speed of retreat of the Niagara River cataract document that the entire Falls are creeping upstream at an average annual rate of about 2½ feet per year. The Canadian Falls section creeps at about 4½ feet per year.

The precipice of the Falls is now very much longer at its ledges than the width of the gorge which it has cut. The flow of water over the ledges is now much shallower, with correspondingly less pressures than existed, say, 3,000 years ago, when the Falls was in the gorge. As a result, the rate of the erosion and undermining of the cataract, and therefore the speed of its retreat, is less now than during the early existence of the gorge. The creeping speed during the creation of the gorge was comparable to the speed of retreat of the Canadian Falls, where the weight, speed, and pressure of the flowing water are more concentrated than the average over the entire Falls.

In 1891 the Commissioners of the State Reservation at Niagara Falls employed Robert S. Woodward to estimate the time required for the creation of the gorge of the Niagara River. A man of unquestioned integrity and superior competence—later to become president of the Carnegie Institution of Washington, D.C.—he reported that less than 8,000 years had been required to cut the gorge of the Niagara River.

By assuming a creeping speed of 4¾ feet per year—one quarter foot per year faster than the upstream movement of the Canadian Falls, to allow for the additional waters now going over the American Falls—we obtain 7,800 years for the approximate life span of the Niagara Gorge. This figure, however, is subject to correction.

When at Lewiston the Falls were approximately 280 feet higher than they are now, and this indicates that the estimated age of the Niagara River is about 7,000 years.

A diminution of forty feet in the perpendicular height of the Falls for every mile that they receded southward is pointed out in a survey made by New York State Geologist James Hall, as recorded by Sir Charles Lyell. Hall states that the southward dip of the rock strata from Lewiston to the Falls is about 25 feet per mile, with the river channel sloping in the opposite direction at the rate of 15 feet per mile. As a result of this change in the height of the Falls, the rates of speed of erosion and cutback of the upstream retreat of the Falls have been variable and not constant. It has depended on the height of the Falls and on the nature of the rocks being cut.

Assuming approximately constant average yearly volume, the force created by this perpendicularly falling water is determined by the height of the fall. The kinetic energy created by the fall increases with the square of the speed of falling. The speed of fall in turn increases with height, through acceleration by gravity, at approximately 32 feet per second for each consecutive second of its fall. The wreckage and erosion of the bottom layers of the precipice take place at a faster rate with a higher fall. This wear and erosion become greater than the erosion at the ledges over which the waters fall. Sections of the cliffs give way and fall to the bottom. It is apparent that the upstream creepage of the Falls was faster when the Falls were higher.

The birth of the Niagara River and the record of the short span of its life history during our present epoch are proof of a recent careen of the globe and a world cataclysm. During the previous epoch the Great Lakes watershed existed in a tropical latitude and drained into the ocean; but it did not drain by way of the Niagara River as it does now. The Niagara River as we know it today did not then exist. It was born, in its present form, with the birth of our epoch.

Embedded sea shells and corals indicate marine formations

Evidence of Careenings of the Globe

in the ten distinct strata of rocks, from Lake Erie to Lake Ontario, through which the Niagara River flows. Ancient beach lines, ridges, and terraces are found at successive levels. At different levels the rocks also have been smoothed, polished, and furrowed by ice. The grooves in the rocks are tell-tale evidence that they were once on the surface, and that they were located in polar regions, where they supported the moving glaciers when this land area was undergoing successive ice ages.

A most careful record of this formation—and one of the first—was made by Sir Charles Lyell, and is recorded in his book *Travels in North America in 1841-42*. These geological phenomena require, for a rational explanation, a careening globe with attendant world cataclysms. Other discoveries of Lyell at Niagara also seem to require changes in the earth's Axis of Figure for their explanation.

He discovered that the northwest cliffs of the whirlpool do not consist of the normal regional rock formation, but are composed of drift, consisting of sand, gravel, loam and boulders, cemented into a conglomerate by carbonate of lime. Since this is a surface layer now, consolidated by lime, it seems to indicate that it had been an ocean bottom for a long period of time, during a previous epoch.

This conglomerate fills an old river bed, now known as St. Davis Valley, which extends northwest from the whirlpool for about three miles, and at its mouth is about two miles wide. The present northern section of the Niagara River, flowing slightly northeast from the whirlpool, is apparently a movement cutoff, established in our present epoch.

The conglomerate filling the St. Davis Valley could not have become deposited during the cataclysm that ended the epoch just previous to the commencement of our own epoch—the latest world Flood—because of the time required for its cementation. It could not have become cemented unless allowed to remain for a long period of time, unassailed by the present rushing waters of the whirlpool. It appears, therefore, to be a formation resulting from a cataclysm that ended one of the earlier epochs.

The boulders it contains are evidence that it underwent an ice age, below an ice cap.

Geologists have also reported important discoveries of two recent river beds at higher levels than the present Niagara River. One river terrace is twelve feet and the other 24 feet above the present level, suggesting that Goat Island, between the American and Canadian Falls, was once under water, with the same fresh-water shells being found there as in the higher river terraces. Both higher terraces extend to the whirlpool. They appear to confirm successive world cataclysms.

The Mississippi River—like the Niagara River—provides a tell-tale geological time scale showing us how long the earth's surface has remained essentially as it is today. The upstream retreat of the Falls of St. Anthony, on the river at Minneapolis, Minnesota, has caused the formation of a gorge between seven and eight miles long and about a quarter of a mile in width. This gorge provides a cutoff from the broad trough of the old Mississippi River bed used during previous epochs of time.

Thanks to information regarding the locations of these Falls provided by early explorers—first Hennepin and later Carver—we know that the Falls receded, up to 1856, at an estimated rate of about five feet per year, and that therefore approximately 8,000 years must have elapsed from the time when the Falls started, at Fort Snelling, to the time they arrived at their present geological location, at the north end of the gorge cut by the Falls.

A correction factor of 10% to 15% must here also be applied in correctly interpreting this time scale; we thus arrive at approximately 7,000 years for the life span of the gorge of St. Anthony's Falls. Consequently the Falls of St. Anthony and Niagara Falls both give us a time scale of about 7,000 years for the duration of our present epoch of time.

The Falls of St. Anthony, when they were located at Fort Snelling, were 110 feet above the present river grade at that point. Now the Falls are about 40 feet high. From these facts it is logical to assume that there was a greater amount of undercutting of the precipice of the Falls when they were higher and that the waters then landed on the base rocks with much greater

Evidence of Careenings of the Globe

force. There was therefore a correspondingly faster upstream creep during the youth of the Falls than during its old age. (The probable profiles of the Falls at various times in the past, with plan and elevations, are shown in *Geological Survey*, Folio 201, Minneapolis–St. Paul, Minn.)

Geological evidence discloses that there were troughs for the bed of the Mississippi River in former epochs of time. One trough now buried commences on the Minnesota River about four miles west of Fort Snelling, circles and joins the present river where the break-through of the present cutoff occurs—just above the Falls of St. Anthony. Here again the river's bed or trough becomes wide and eroded.

The Minnesota River occupies an oversize and eroded trough which continues beyond its source, crosses the continental divide, and is continuous with the channel of the Red River of the North which now flows in the opposite direction.

These geological features show that a change of land elevation occurred in this area about 7,000 years ago. The old river troughs were cut in previous epochs. The narrow gorge or cutoff was created during the present epoch.

The Mississippi River's delta contains a record of the river's age. The number of cubic feet of sediment, and the number of millions of tons of earth materials being carried southward by the waters of the river and deposited in the Gulf of Mexico, have been determined with fair accuracy.

Based on these data and the dimensions of the delta of the river, a close estimate of the age of the delta can be made; this will also give us the age of the present river. But estimates of the age of the river—already made by those whose opinions are most highly regarded—vary from 4,000 years to 138,393 years (*Geological Bulletin* #8, Louisiana Department of Conservation). This wide variation in expert opinion is due to the total lack of agreement as to which of the many substrata of the river bed are to be considered as belonging to the delta.

The head of the delta has been found, by different individuals and groups, to be at various places—including Keokuk, St. Louis, Cape Girardeau, Commerce, the mouth of the Red River, and Baton Rouge. Each of those places is probably a correct location for one of the former delta heads, for there have been many.

The general acceptance of the new theory of a recurrently careening globe will result in resolving all differences of opinion as to the location of the head of the delta of the present Mississippi River. Each time the earth careened a new delta head became established, for the watershed existed as far back in geological time as we can go.

During Epoch No. 1 B.P., preceding the latest careen of the globe, the Mississippi River flowed from west to east in a tropical climate. The Sudan Basin, now in Africa, was then at the North Pole. In Epoch No. 2 B.P. the river flowed generally southward, in a frigid climate. The Hudson Bay Basin land area was at the North Pole and in summers glaciers fed the upper river.

Cores from borings have been taken in abundance by the Mississippi River Commission to establish the approximate areas and depths of the successive alluvial strata and to determine the various channels of the old river in past ages. Variations in the character of the fills and a study of the diatoms, algae, foraminifera, and other fossil evidences will disclose changes of climate.

Remains of tropical vegetation and water life should mark the next delta layer below the surface strata. Next below that should be evidences of the flow-off of the glaciers which melted at the headwaters of the river at the very beginning of Epoch No. 1 B.P. Below this, the evidences should disclose a cold-water river whose northern headwaters were fed by the North Pole Hudson Bay Basin glaciers during Epoch No. 2 B.P.

To determine the age of the present river it is necessary to identify the area and volume of the present top delta and divide the weight of its cubical contents by the weight of the average annual deposits of sediment.

The following table results from a preliminary effort to tie together epochs and strata that are now generally recognized.

Epoch	Duration Years Approximate	U.S. Geological Survey Series of Epochs	Mississippi River Commission Designations for Substrata
Present	7,000		Recent Alluvium
1 B.P.	4,400		Jackson
2 B.P.	7,000	Wisconsin Ice Age	Vicksburg-Jackson
3 B.P.	5,000	Peorian Life Age	Claiborn
4 B.P.	—	Iowan Ice Age	Wilcox
5 B.P.	—	Sangamon Life Age	Midway
6 B.P.	—	Illinoisan Ice Age	Upper Cretaceous
7 B.P.	—	Yarmouth Life Age	Lower Cretaceous
8 B.P.	—	Kaasan Ice Age	Mesozoic
9 B.P.	—	Aftonian Life Age	—
10 B.P.	—	Albertan Ice Age	—

Great Salt Lake is a shrinking remnant of a much greater lake—known as Lake Bonneville—which in prehistoric times filled the entire present-day land basin.

The Bonneville shore level, wave-cut by the former lake, shows it to have covered an area of about 20,000 square miles, with a depth of 1,050 feet. (*Bulletin of the University of Utah*, Vol. 30, October 1939, No. 4). The area of the existing lake is about 2,000 square miles, which is about the size of the land area of the state of Delaware.

Proofs of the recurrent careenings of the globe have been developed from the original proposition that the continents of North and South America lay along the equator, in tandem, during the epoch of time just preceding the epoch in which we are now living. This theory receives confirmation from the old shore lines, beaches, and wave-cut terraces in the rocks—now high above the surface of Great Salt Lake.

When the present lake basin was located in an equatorial area, it received as great a rainfall as now prevails along the Amazon River. Such torrential rainfall can be considered sufficient to fill the lake basin to the Bonneville shore line level, since lake levels are maintained by the balance between rainfall on the watershed and the rate of evaporation, or—where there is an overflow outlet—by its elevation.

Descriptions of soil borings confirm the lake basin's tropical location in Epoch No. 1 B.P., and also its successive locations in tropical and nontropical latitudes. Excellent illustrations of the many different prehistoric shore lines and other features appear in the comprehensive report on Lake Bonneville by G. K. Gilbert in U.S. Geological Survey, Monograph #1, 1890.

Ice Ages

LOUIS AGASSIZ, around 1845, first used the phrase *ice age* to account for glacial markings on rocks. Since then many persons have assumed, erroneously, that the glacial markings indicated a change in climate throughout the world. The term Ice Age is

Evidence of Careenings of the Globe

now used to define a relatively small area, approximately within a circle, such as the area now contained within the Antarctic Circle. We are now living with what may be called "The Antarctica Ice Age." A lesser Ice Age includes those areas of Greenland, North America, and Asia that lie within the Arctic Circle.

Ice Ages have occurred in all the continents of the world, as indicated by the tell-tale scouring marks on successive layers of rocks. These rock groovings always radiate from central points which indicate the locations of the North Pole or South Pole of that particular epoch of time. The surrounding polar areas never received enough heat from the sun to melt the ice which accumulated from the constant snowfall.

During each successive polar Ice Age the rest of the globe enjoyed tropical or temperate climates, as at present. We know this because the fossils of animal and plant life indicate the climates in which each section of successive earth strata existed and they tell us clearly that the globe has rotated on many successive Axes of Figure.

Five successive Ice Ages have left their scars in land areas of Canada and northeastern United States. Glacial markings on rocks, loose boulders and debris, are in evidence over most of this area.

Life Ages have occurred in these same land areas between the Ice Ages. These Life Ages were long intervals of time during which these regions were free of glaciers, and were warmer than at present. Each Ice Age blotted out the Life Age of a certain area and was, in turn, succeeded by another Life Age in the same region. These changes were sudden and without gradation.

The sudden birth of each Ice Age was paralyzing and destructive to animal and plant life. Each Ice Age produced an ice cap which grew to maturity and by its great weight depressed and dented the earth beneath. The ice caps uprooted and carried on, under and within their massive slowly moving bodies, enormous sections of rock and earth to be deposited elsewhere; while, beneath the ice, the land was gouged, earth and rocks were scattered all around, and valleys and river beds were filled with debris.

The end of the ice caps did not come through slow withdrawal. When they left, they disappeared just as rapidly as such huge masses of ice can melt when moved to a tropical climate.

Evidences of the Life Ages are found under and above the successive overlapping flows of till and glacial debris which have been carefully charted by the United States Geological Survey. During each Life Age animals and plants lived and multiplied, forests flourished, bogs developed, brooks and rivers flowed, valleys filled, while mountains and rocks eroded.

Fossil remains of animal and plant life are absent from the strata of till and debris representing the five Ice Ages occurring between the Life Ages. Absent also from these strata are the marks of erosion of soil and rock and the usual evidences of the normal work of rivers and brooks in filling valleys with transported sediment. The waters had been changed to solid ice. Glacial debris is the sole remaining evidence.

Today, those land areas are in a part of the globe which is having a Life Age, while the land areas of Antarctica and Greenland are passing through a temporary Ice Age. Canada and northeastern United States are now in a second successive Life Age, since they once were covered by an ice cap, or formed a frozen tundra adjacent to it.

In the area now known as New York State several whole mastodons have been discovered, as well as parts of over two hundred mastodons and of about fifteen mammoths. Like the mammoths found in Siberia and Alaska, the whole mastodons were perfectly preserved after their death by quick freezing and cold storage during the first Ice Age. Also as in the case of the mammoths, they have been found with full stomachs, indicating the sudden death of a healthy animal.

These mastodons lived in a warm climate, in Epoch No. 3 B.P., perished at the close of this period, and were buried in a flood or by hurricane debris—all of which froze solid. During Epoch No. 2 B.P. they were interred in cold storage, the New

Evidence of Careenings of the Globe

York State area then being frozen tundra, and the Hudson Bay area being at the North Pole.

Subsequently, the North Pole was in the Sudan Basin area (during the following epoch, No. 1 B.P.), and the New York State area—with its interred mastodons—was moved to the tropics. Both soil and dead animals immediately thawed out; but the soil which preserved these particular specimens must have been so moist that oxidation was retarded.

Now they are in a temperate climate, during a second Life Age for this land area, and are frequently found in a collapsed and disintegrating condition.

There is much geological evidence to show that the last Ice Age of the North American continent was caused by a great ice cap centered in Canada that extended southward to present-day New York City. Moraines, eskers, clay beds, and other residual evidence of the glacier exist in many places, including northern New Jersey and southern New York.

A shallow dent in the earth, averaging 420 feet below sea level, now known as Hudson Bay, marks the ice cap's approximate center, while the heights of land known as the Laurentian Shield and which almost surround Hudson Bay, mark the final edges or lips of the main ice bowl.

The watershed of the Hudson Bay Basin corresponds to the kind of scar or dent in the surface of the earth which an ice cap could make, and which it would leave behind as evidence of its existence. Counts of the annual varves (layers) in clay beds at New London, Wisconsin, and Hackensack, N.J., indicate that the Hudson Bay Ice Age lasted for approximately 7,000 years!

The distance from the North Pole of Figure of the epoch, in Hudson Bay, to the moraine of Long Island is approximately 1,800 miles. Analytically, this compares with the distance from the present South Pole of Figure to the ocean, this being about 1,800 miles for most of the perimeter of the Antarctic Continent. At the Ross Sea the distance is about 600 miles.

The flow-offs of glacial ice, during the Hudson Bay Ice Age, were distributed in much the same way as the present flow-offs of glacial ice in Antarctica. The ice flowed until it reached the ocean, flowing faster in the direction of Hudson Strait and Davis Strait than toward Long Island, N.Y., because the ocean was nearer in that direction and the grade was correspondingly steeper; it accumulated greater volumes of ice where the flow-off was most retarded.

In the southern region the grooves in the rocks show that the ice was flowing south, but in the northern regions the markings show that the ice was flowing north. Glacial striae caused by the movements of this last North American ice cap, together with displaced boulders, are in evidence as far south as Pennsylvania, Ohio, and the Mississippi valley, and as far north as the Northwest Territory of Canada.

The outstanding fact is that the ice radiated from a center in the area now known as the Hudson Bay Basin. It did not spread southward from the present North Pole area. The ice flowed away from a central point; this shows clearly that the Hudson Bay Basin was then at the North Pole of the Axis of Spin, and that the ice flowed in every direction from the pole.

In Antarctica today the very same types of striations are being created on the rocks, and boulders are continuously carried seaward (from the South Pole) on glacial ice flowing northward.

Over a hundred years ago Louis Agassiz discovered glacial markings in the Amazon Valley, along the Equator! On both sides of the Equator—within 18 and 20 degrees—glacial markings have been found in Permian rocks. In other regions of the world, tree specimens with annual rings have been found in rock formations of the same age indicating that temperate zone conditions prevailed in the regions where the trees grew, at the very same time that polar region glaciers scoured and striated the rocks—at that time located near the poles but now being near the Equator.

Ice Ages are recorded in rocks at random latitudes and longitudes for all periods of geological history. For example, in all continents glacial horizons are found in rocks classified in geological textbooks as Pre-Cambrian and Permian. Three of the Pre-Cambrian locations are in Africa, three in Asia, and two in Australia. Five of the Permian glaciated horizons are in South

Evidence of Careenings of the Globe 49

America. Five of the most recent Ice Ages are recorded in the rocks of Canada and the United States.

In all cases where the records can be studied, it has been found that the striae show radiations of the scouring ice masses from central points. This gives adequate evidence in support of the theory that the glaciated areas were at the poles of the earth in the epoch of time during which each ice cap was developed. They moved away from the poles when they reached maturity and caused the globe to careen.

Geological Outcroppings

THE age of various earth strata are determined by studying the fossils found in them, and generally the lower strata correspond to the earlier epochs of the earth's history—but not always. Outcroppings of very old rocks appear in many places on the earth's surface instead of being buried deep in its bowels. The "Old Red," for example, a very hard and durable sandstone, which is classified as belonging to the Lower Cambrian Period, is found as surface outcroppings in New York State, in West Virginia, and Canada. It should normally be buried at depths of three miles or more, according to the geology charts.

Such facts indicate that the slow building up of the earth, epoch by epoch, layer by layer, through ages of time, has not been an uninterrupted process. A mighty cataclysm has taken away all the overlying rock strata and the earth materials above the hard red sandstone in the areas where the stone now appears as an outcropping, but that same cataclysmic force was not on the loose in those other areas where the overlying materials are still intact.

The gouging out and tearing away of earth strata to a very great depth in certain areas have been caused by such materials coming into contact, at high speeds, with masses of ocean waters churned into a swirling flood during the careens of the globe.

This operational force of Nature is the logical reason why the "Old Red" sandstones are being found as outcroppings on the surface of the earth; isostasy accounts for the earth again being rounded off following the new arrangements of the earth's masses of land and sea.

According to older authorities, the transporting power of water equals the 6th power of its velocity. It is given as 3.2 to the 4th power (3.2^4) by P. G. Worcester in his recent textbook in geomorphology. He states that wher the St. Francis Dam in California failed in 1928, blocks of concrete weighing more than 10,000 tons were carried more than half a mile down the valley.

During the transient cataclysms caused by land masses careening against the ocean waters, the pressures created at maximum speeds of careening are beyond the imagination of man. Vegetation is crushed to a pulp and animals are obliterated.

The ancient earth materials which once covered the "Old Red" sandstone, in successive layers, and which were torn loose and washed away by one of the great deluges, are now superimposed somewhere on more recent earth strata, and may be either heaped in ridges or blocks or spread widely in conglomeratic strata.

In Cuvier's "Essay on the Theory of the Earth" it is stated that "Mr. Kerwin has given weighty reasons for his belief that the globe has been, at some remote period, most violently assailed by a mighty flood from the southeast. Tearing up and bearing away the looser materials of the southern hemisphere, it has brought a great body of them to the northern, and impressed upon the capes of Good Hope, of Horn, and Van Diemen's Land (Tasmania), and other promontories, the marks of its overwhelming force."

There is geological evidence that mountains have been cut off and carried away during the cataclysms of the deluges. Miles of telltale slanting rocks exist in normal formations which appear to have been cleanly cut off. A fairly level plain is now all that remains where once a mountain stood at Joggins, Nova Scotia, on the Bay of Fundy, where the fossil forests are also on display.

Martin Gardner, referring to Chief Mountain in the Alberta-Montana region of the Rocky Mountains, where older strata are found resting on younger, states that "In its form the fault

Evidence of Careenings of the Globe

line of the overthrust can be seen clearly, with slickened faces of rock which testify to the faulting movement." He mentions Hart Mountain in Wyoming as another upside-down spot, and he states, "The fault line is easily traceable for some twenty-five miles." (*Facts and Fallacies in the Name of Science*, Page 130.)

Table Mountain, at Capetown, South Africa, rises 3,500 feet above sea level. It consists of horizontal layers of sedimentary rocks still intact. All of the surrounding layers of rock were gouged out and washed away by the impact of the ocean waters when the southern tip of the continent of Africa spearheaded the southward careen of that continent during the latest great deluge.

The gouged-out profile of Table Mountain and other land strata of the southern tip of Africa illustrate the effects of some of the forces of nature which were active when that continent careened southward into, under, and through the inert ocean. The oceans around Africa became extremely turbulent because of the sudden change of ocean depths with change of latitude, and because of the motion of the great land mass which careened into them. It was the reverse of an ordinary flood or of the overflow of a mighty river. In this flood the land moved against the waters, and the waters then rode up over some of the land.

The Koran confirms the mechanical force of the flooding waters: "The earth's surface boiled (seethed, roiled) up . . . the Ark moved . . . amid waves like mountains."

It was upon these temporarily turbulent waters, near the east coast of Africa, that the vessels of Noah and Deucalion rode this latest flood; the land below them careened southward, until Noah's barge bumped or was bumped by the summit of Mt. Ararat, and Deucalion's life-saving chest ran aground on Mt. Parnassus.

There are remnants of tablelands in a great many places throughout the world; here layers of sedimentary rocks, which previously have been below the surface of the earth are at the surface in the form of gouged-out ends or sides of mountains, buttes and hills, exposing great numbers of the successively created horizontal stratifications, all cut off cleanly, stripped from top to bottom. Typical examples can be seen at Monument Canyon, Arizona, in Yellowstone National Park, in South Dakota and western Nebraska, near Banff in the Canadian Rockies, near New Haven, Connecticut, and at the Delaware Water Gap. Most of these cutaway mountains and buttes represent what was left when the waters of a great deluge tore away all the land excepting the gouged sections still standing intact as mountains. The exposure of the inner layers of these ravaged mountains is indicative of the force of the impact of land and water during the cataclysms produced by the careenings of the globe.

Another reason for the occurrence of older rock formations on top of younger is the transportation work of glaciers. The compound word "Ice-Age" created only about 100 years ago, and the ice-age theory, suggested at that time, enabled puzzled geologists to account for the appearance of "foreign" rock materials in many places, and for many of the unconformities in successive earth strata.

A few miles south of Lake Erie, near Jamestown, N.Y., there is a huge erratic block of whitish stone perched on the summit of a small mountain range of brownish stone. It marks the spot it had reached when the glacier it was riding turned to water. It settled down like a Noah's Ark.

The copper-bearing rock formations of Arlington and Kingsland, N.J., are not native, but are erratic blocks, transported by ice from some—as yet unidentified—region, and dropped when the ice melted. There are many similar erratic blocks in many parts of this and other countries.

Long Island, N.Y., was created by glacial action, being a terminal moraine in its north sections and an outwash plain in the south, formed by the heavy loads of silt, sand, and gravel carried by the streams emerging from beneath the last North American glacier. In Boston Harbor the islands are composed of glacial till and their long axes are parallel to the direction of the flow of the ice. They are called drumlins, and were left by glaciers.

When the great size and carrying capacity of the recent consecutive North American polar glaciers are analysed and fully appreciated, many other unconformities of earth, rocks and small mountains will be adjudged to have been glacially transported to their present locations.

The Oldest Rocks and the Age of the Earth

THE oldest rocks about which we can have any definite knowledge are those within our physical reach. We know well the rocks from various earth strata that have been obtained from mine shafts, tunnelings, and well drillings. We also know well rocks which have formerly been buried, perhaps three miles below the surface, but are now exposed as outcroppings. But what we know about other rocks must remain hypothetical—being in the nature of philosophic geology.

Nevertheless, we do possess much valuable information. As tentative as our calculations and theories must be, they are supported by substantial evidence which serves as a basis for initial theorizing and as a guide for future research.

A legible record of the frequent careenings of our globe, with specimens of animal and plant life of each epoch, is contained in the great stone book whose pages are the successive strata of the earth's materials—thousands of which have been penetrated and have been sampled and identified on the basis of cores from drillings.

Men have explored beneath the surface of the earth to a depth of approximately five miles by boring in search of oil, and to lesser depths in search of water and other minerals. The evidence brought up in the drilling cores proves that the earth is built up of layers. There is no reason why the strata of the earth below those penetrated by the deepest well drillings should have been created in a different manner from the strata above.

From the depth of the strata encountered in the deepest drilling to date, their average thickness, and the total number of layers, we can estimate the age of the rocks penetrated. On the basis of data indicating the ultimate depth of the earth's stratifications, we can estimate the age of the earth.

The deepest drilling on record penetrated 25,340 feet below the surface of the earth. It was drilled in 1913 by the Phillips Petroleum Company in the Pecos Field in Texas. The average strata of the earth through which it penetrated, as shown in the Spur Ranch Drilling data of the next chapter, were approximately 13 feet in thickness, and each stratum was created during an epoch lasting approximately 6,000 years. Hence, 25,340 feet divided by 13 feet per layer yields the number of epochs transversed by the drilling—1,949. This figure multiplied by 6,000, the average duration of each of the recent epochs, gives the approximate age of the rock stratum at the bottom of the well—11,694,000 years.

The oldest rocks about which we have any accurate knowledge are therefore around twelve million years old. Factual evidence of rocks that are older is not available at present.

A further analysis of the age of rocks, based on the new theory of continuous creation and the constant buildup of materials of the earth, is given in Part III, "The Origin of the Earth's Materials."

There is no known reason why the strata below those penetrated by the deepest well drillings should have been created in a different manner from the strata above. Thus, by determining the ultimate depth of rock strata in the earth, we can estimate the age of the earth in roughly the same manner that we determined the approximate age of the oldest rocks we know. This is done with the aid of depth soundings; by artificially creating shock waves through the earth, akin to the shock waves of earthquakes, we can measure the depth of the rock strata.

The estimated depth of solid materials which exist and transmit seismic waves between the core of the earth and ground level are discussed in *Encyclopaedia Britannica* and in *Physics of the Earth*, Vol. VII, published by the National Research Council. Terrestrial depth soundings have indicated that there is something about 1,850 miles below ground level from which the

Evidence of Careenings of the Globe

impulse waves bounce back to the surface. This is assumed to be the core of the earth.

In order to determine fairly exactly the age of the earth, however, the figures used so far must be adjusted.

We have said that presently available data show that the upper strata of earth materials each required about 6,000 years for their creation. This is the estimated duration of each epoch during which the upper layers were formed. However, it seems logical to assume that the time periods between careens of the globe were shorter when the globe was smaller. Thus, 4,500 years between careens of the globe are assumed for this tentative calculation of the earth's age, and the rate of buildup of earth materials is set at 9¾ feet per epoch instead of 13 feet.

According to these estimated figures, then, the number of feet from the surface of the earth to its core (1,850 x 5,280) divided by the number of feet per stratum (9¾) indicates that the impulse waves travel through one million strata to reach the core. Taking 4,500 years as the duration of each epoch, we find the age of the earth to be approximately 4½ billion years.

As suggested throughout, these figures for the age of the earth and the age of the oldest rocks known are tentative. They are based solely on the Spur Ranch Drilling, but this one was more accurately supervised than the many commercial drillings. Many similarly supervised drillings in numerous areas the world over must be made and compared to enable one to make more perfect estimates. More accurate data are also required regarding the duration of the successive epochs.

Epochs of Geological Stratification

THE word "epoch" as here used denotes the period of time, during which the globe rotated on any one axis of figure. The careening globe theory is supported by evidence to the effect that most of the land areas of the earth have changed their latitudes, and that major changes in the arrangement of the materials of the earth's surface occurred in, and marked the end of, each of these epochs.

The theory is one of normal and natural cataclysmic changes occurring at the end of each epoch, and of other normal and natural, but slower, changes being continuously wrought during each epoch, both types of change being the natural results of definitely identified operational and creative forces of nature.

The history of the earth—as written in the rocks—indicates that both kinds of change have taken place. Fossil evidence proves that many different kinds of animals, reptiles, fishes, shell-fish and plants lived for long periods of time and then ceased to exist.

The fossil remains of extinct forms of life—now buried in the depth of the earth—show that they once lived on the surface of the earth or in surface waters. They are found in the oldest rocks known, and are found in cores brought up by the deepest oil-well borings. They show us that the creation and development of animal and plant life, and of earth materials, in the upper five miles of the earth's materials have been gradual and continuous for the whole earth.

The most recent epochs can be counted by an examination of the upper layers of the earth's materials—much as the age of a tree may be determined by counting the number of rings on its stump; and, just as the climate prevailing during each year of the tree's growth can be learned from the condition of each ring, we can analyse the condition of each of the earth's layers. There is a difference, however; while a tree has one ring per year, an epoch may be represented by several layers in any one local area, or some of the strata may be missing.

Numerous and various noncomformities will be found— varying with the geographic location of each area during each epoch. For example: a local stratum may be developed from organic life, from vegetation and animals, together with dust and dirt carried by wind and water; this soil may then be all or partly washed away by the waters of one of the world deluges, or may become covered by sand and flood detritus or by the glacial drift of an ice age.

At this point in our examination of the evidence for the

Evidence of Careenings of the Globe

careening of the globe we can piece together the materials we do have to describe the recent geological epochs.

The age of the epoch in which we now live is tentatively estimated at 7,000 years. It will be a useful yardstick, especially when comparing the length of our epoch with the life span of earlier epochs. However, it may be adjusted when better evidence becomes available.

A period of 7,000 years conforms approximately to the historical period, beginning at or about 5000 B.C., and also to the time scales of the gorges produced by the Niagara and Mississippi Rivers.

Materials containing Carbon 14, properly identified as having been contemporaneous with the last great cataclysm of the earth, can be counted on to produce close estimates of the elapsed time since that cataclysm occurred. Carbon 14 datings of mammoths, rhinoceroses, mammoth trees, and other vegetable and animal remains buried in the tundra of Siberia and elsewhere at the time of that cataclysm, may disclose ages of less than 7,000 years. The more data we obtain on this subject the shorter the elapsed time seems to become.

The idea occurred to me that the perfectly preserved mammoths found in the Arctic tundra had possibly been reeled from a tropic to a frigid zone; and that, if so, perhaps I could find some evidence to that effect.

I imagined having an 8-inch globe of the earth in my hands, then throwing it into the air, and making it stay there and spin around. I thought that North and South America would ride along the Equator of Spin, due to their weight and the corresponding centrifugal force, provided I kept the globe spinning.

It seemed common sense to say that the earth would naturally rotate with its heaviest masses stretched out along the Equator, because the centrifugal force throws the greatest weights to the periphery or equator of any freely suspended rotating mass. And yet, quite to the contrary, the mighty Rocky Mountains and Andes Mountains are not now stretched along the Equator.

My theory was that if I could plot a circumference of the earth which might have been the Equator when the mammoths were living—just prior to the latest careen of the globe—then I might be able to find some evidence of a polar area 90 degrees of latitude away from that Equator, and this would prove the theory to be correct.

I was curious enough to make a trip one Saturday afternoon in 1913 to the Public Library of Erie, Pa., where I then lived. I took along a spool of ⅛-inch-wide red ribbon, and tied the ribbon around a three-foot globe of the earth, which stood in the middle of the main library room. I remember that I felt self-conscious at first, but soon guessed that the other library patrons must think that I was just one of the men who worked there, for they paid no attention to me, and I worked leisurely. What was routine at the time, now in retrospect looms as a momentous occasion.

I tied the first ribbon in a great meridian circle, or Equator-like line, along the great ridges of the Rocky Mountains and the Andes Mountains. It divided the globe into two equal halves. On the opposite side of the globe, I was interested to note, the ribbon traversed East Asia and the shallow seas surrounding the Malay Peninsula. I tried to make it represent the meridian or equatorial band traversing the heaviest land areas, and it seemed that the land zones it touched far outweighed any other circumferential belt of the earth's surface that I could have selected.

Then I attached two other ribbons—representing great circles of the earth—at random places, but taking particular pains that they were at exactly a 90-degree angle to the first band. The idea was that I might find some evidences of former polar areas where these two upright ribbons intersected each other, for the intersections would mark the locations of polar centers, at the moment of time that my first ribbon was an actual under-the-sun equator.

One intersection was found to be at Lake Chad, Africa—which I thought might give me a clue; the other intersection was located in the Pacific Ocean, and no clue came to mind. I recall

Evidence of Careenings of the Globe

writing "Lake Chad" on a piece of paper, with a resolve to try to find out something about it. I was looking for evidence of glacial action, or of a dent left in the earth by an ice cap. I was astonished to discover just what I was looking for: the great Sudan Basin of Africa.

"In North Africa there is a vast space of upwards of four million square miles, extending from the Nile valley westward to the Atlantic coast, and from the plateau of Barbary in the north to the extremities of the basin of Lake Chad in the south, from which no single river finds its way to the sea. The whole of this space, however, appears to be furrowed by water channels in the most varied directions.

"From the inner slopes of the plateau of Barbary numerous wadys take a direction toward the great sand belt of the Erg, in which they terminate; a great series of channels appear to radiate from the higher portion of the Sahara, which lies immediately north of the tropic of cancer and in about 5° E. of Greenwich; another cluster radiates from the mountains of Tibesti, in the eastern Sahara." *Encyclopaedia Britannica*, 9th edition, Vol. I, page 255.

This formation could be the natural result of the torrential run-off of melting glacial ice, turned to water by a blazing tropical sun, when the Sudan Basin Ice Cap reeled from the North Pole of Spin and melted near the Equator, about 7,000 years ago. The innumerable watercourses—without any apparent relation to one another—remain as tell-tale evidence.

Lake Chad has one outstanding peculiarity. It is a freshwater lake, and yet it does not have an outlet to the sea. There is not another lake like it on the surface of the globe.

It is generally understood that lake water remains fresh by surrendering its salts to the oceans, through run-offs, or overflows. With the exception of Lake Chad, the large lakes without outlets to the sea are salty. They were filled with sea water when first created at the beginning of the present epoch. Some—such as the Caspian Sea—probably overflowed and lost their salt contents due to the excessive humidity created by the melting of the Sudan Basin Ice Cap and the eastward-flowing air currents, but these lakes later shrank, lost their overflows, and have since accumulated mineral salts.

Salts are carried and brought in by the incoming tributary rivers and streams. As indicated before, the water levels of lakes are determined by the ratio of incoming water to evaporation. During the approximately 7,000 years of its existence Lake Chad has not accumulated sufficient salts from its tributaries to cause it to become salty. It remains a fresh-water lake without an outlet to the sea.

During the rainy season Lake Chad overflows its normal basin. The *Encyclopaedia Britannica* statement adds that it fertilizes the wadys—including the great wady or basin to the northeast, into which it overflows. This indicates that the overflow does not purge the lake of salts to any extent, because if it did, the soil in the overflowed areas would become salty and not fertilized.

The Equator of the last epoch of time prior to our own was a line along the Rocky and Andes Mountains. If the earth careened 80 degrees of latitude, instead of 90 degrees, then Lake Chad would not be the polar center of the previous North Pole Ice Cap. Lake Chad is located several hundred miles southeast of the center of the Sudan Basin.

The Sudan Basin is in the center of the north lobe of that continent. The accumulating ice mass, which created the great depression of that basin, was landlocked and therefore it did not have a chance to flow off the land into the sea through the force of gravity, in the manner that the glacial ice now flows off the smaller continent of Antarctica. This fact, together with the probability that a smaller ice cap grew at the South Pole of that epoch, would appear to account for Epoch No. 1 B.P. having had a shorter life span than the epochs just preceding and following.

Further studies of the channels cut by the watercourses in the Sudan Basin and of the glacial striations cut on Canadian rocks may be expected to show that neither Lake Chad nor Hudson Bay were the exact centers of the North Pole Ice Caps of those two epochs. But, Lake Chad and Hudson Bay give us two points of reference; they have moved approximately equal distances from the Pole of Spin each time that the earth has careened and are therefore useful references to the approximate polar points of their epochs.

The Hudson Bay Ice Cap has left a more sharply marked, and larger, dent in the earth than has the subsequent Sudan Basin Ice Cap. The Hudson Bay Ice Cap took a roughly 60% longer time to grow and therefore made a larger and deeper basin than the Sudan Basin Ice Cap. The process of isostasy, tending to raise sunken areas, adds to the difficulty in reading the records correctly.

When the Sudan Basin Ice Cap moved away from the North Pole of Spin, it did not go all the way to the Equator, because by the time it reached its present position, the new bulge of the earth had become established. The Sudan Basin Ice Cap and the new bulge of the earth were soon traveling along together, at the rotation speed of the earth—or about 1,000 miles per hour.

The careening speed of the Ice Cap had practically vanished and its eccentric pull of centrifugal force had now been changed from a sideways pull to an upward and outward stabilizing force. Isostatic equilibrium had once again become established, subject to readjustment through earthquakes, until the Ice Cap had completely melted.

Thus Epoch No. 1 B.P. ended at the time of the latest world flood, and the present epoch commenced. During Epoch No. 1 B.P. the Sudan Basin area of Africa was occupying the area at the North Pole of Spin. The continents of North and South America then lay in tandem on one side of the globe, along the Equator, and the eastern parts of Siberia and China were on the opposite side.

The time estimate suggested is based on recent advances in pinpointing geological time by measuring the amount of Carbon 14 remaining in sample specimens of wood. Wood from an ancient forest, uncovered in a lower stratum of the earth at Two Rivers, Wisconsin—geologically tied in with the ending of the latest North American Ice Cap, often referred to as the Wisconsin Ice Age—was analysed for its Carbon 14 content.

Its age was found to be approximately 11,400 years old. The literature dealing with this ancient Wisconsin forest contains many reports on its age, varying from 11,000 to 12,000 years. I have chosen to use the report by the foremost authority in this field, Willard F. Libby, who arrives at a figure of 11,400 years.

If we subtract 7,000 years, the estimated age of our present epoch, from 11,400 years, we are left with 4,400 years as the duration of Epoch No. 1 B.P. The figure is subject to correction when better data become available.

During Epoch No. 2 B.P. the geological dent of the Hudson Bay watershed contained the North Pole and its ice cap, while South America, Africa, Borneo, and India lay along the Equator. During that epoch clay beds were laid down by silt, presumably carried by glacial streams flowing in summer from beneath the edges of the Hudson Bay Ice Cap. Each layer of clay—called a varve—corresponds to a single year's growth of the clay beds; thus the approximate duration of that epoch may be set at 7,000 years.

The identification of this former North Pole area is based on the fact that its distance from Lake Chad in the Sudan Basin of Africa is approximately the same as the present distance of Lake Chad from the North Pole, which, in turn, is the distance through which the Lake Chad area moved during the latest careen of the globe. All careening moves cover about 80 degrees of latitude.

The depression now occupied by the Caspian Sea seems to have been the location of the North Pole during Epoch No. 3 B.P., because its bowl—as well as that of the Black Sea—was formed by a depression caused by an ice cap.

The Caspian Sea is located in a large low-lying land area—called the Caspian Depression—and is the focal drainage center which gathers the gravitational downhill water flow of several large rivers, including the Ural and Volga rivers. This sunken Caspian area has geological similarities to the depressed areas in which Hudson Bay in Canada and Lake Chad in Africa are located. All three are drainage centers for extensive river systems

Evidence of Careenings of the Globe

in very large areas, and have been depressed into basin formations by the weight of the ice caps of former epochs.

We know that each successive ice cap leaves a depression in the land, and each careen of the globe moves both the North and South Pole ice caps distances of about 80 degrees of latitude—around 5,500 miles. With this information as a guide, we can identify the locations of former poles of many successive epochs. We find approximately the same distances between the Caspian Sea and Hudson Bay, between Hudson Bay and Lake Chad, between Lake Chad and the present North Pole.

The one observable clue to the duration of the Caspian Basin epoch—No. 3 B.P.—is that it did not permit a ready flow-off of the ice to the oceans. It appears to have been landlocked. Therefore, it is assumed to have had a shorter life span than the Hudson Bay Basin epoch—No. 2 B.P. It was more like the Sudan Basin Ice Cap of Epoch No. 1 B.P. Possibly, a duration of 5,000 years is a fair estimate.

The recent geological history of the earth—subject to being corrected when better data become available—may be summarized as follows:

Epoch No. 3 B.P. began about 23,400 years ago and lasted around 5,000 years. The Caspian Depression was at the North Pole of Spin. The New York State mastodons were living in a tropical climate.

Epoch No. 2 B.P. lasted around 7,000 years. The Hudson Bay was at the North Pole. The New York State mastodons were in cold storage—like the Siberian mammoths of today.

Epoch No. 1 B.P. lasted around 4,400 years. The Sudan Basin of Africa was at the North Pole. The New York State mastodons were buried in the tropics. The Siberian mammoths were grazing in an Eden-like climate in or near the tropics.

Our own Epoch has already lasted about 7,000 years. The Arctic Ocean is at the North Pole. The mammoths are buried in frozen tundra in Siberia and Alaska.

The mystery of how the earth was created and built up can

now be explained through the rational and understandable display of factual evidence. We now know just how the various layers of rock, sand, shale, etc.—with their telltale specimens of former animal and plant life—came to be laid down in the strata in which they are now found. For example: A piece of live fresh wood—resembling red cedar—was found during the excavation made for the foundation of the Chase Manhattan Bank building at 18 Pine Street, New York City. It was found three feet below the top surface of a layer of hardpan, which in turn was about sixty feet below the surface. The tree probably grew in Epoch No. 2 B.P.

In the excavation made for the foundation of the New York Telephone Company's building at Barclay and West Streets, New York City, the contractors came across the prostrate trunks of several juniper trees—with bark and branches intact. At 45 feet below high tide level they came across a bed of peat eighteen inches thick. The peat bed had grown where found; probably the trees also. These might tentatively be classed as growths of Epoch No. 1 B.P.

Parts of whale skeletons have been found on Long Island, N.Y. These are tentatively classed as belonging to our present epoch, because during this period of time the ocean waters are assumed to have receded and now make up part of the ice and snow in the Antarctic Ice Cap; also, the dearth of top soil on Long Island lends support to the idea that it has not had its present size during the approximately 7,000 years of our epoch.

Clay beds abound in the vicinity of New York City; they are found with layers of sand, gravel, loam or topsoil, etc., superimposed upon them. Facts such as these agree with the theory of the careenings of the globe.

By assuming that there was land at the edge of the Hudson Bay Ice Cap, during Epoch No. 2 B.P., it follows that silt from the glacial streams settled in lakes or rivers and formed these clay beds. Since then, two world deluges and two epochs of time have occurred, during which the layers of overlying materials now being found were developed.

Clay beds are normally laid down in layers of silt with dif-

Evidence of Careenings of the Globe

ferent colors and textures. These varves contain a record of each year; in some instances they indicate the climate, rainfall, and vegetation, by the color, thickness, and texture of each varve.

A physical time scale for our present epoch has been developed from a study of varied clays by the Swedish scientist Gerard De Geer. He fixed the beginning of the post-glacial epoch at about 8,700 years ago. He counted the number of varves in clay beds in river valleys running north-south, assuming a gradual ascension, like a long staircase, as the ice retreated northward, and tying in the top varve of one clay bed with a lower varve of a higher bed. He called it the Swedish Time Scale.

De Geer reported the following numbers of clay varves from locations presumed to have been at the Ice Cap's edges:

Location	Count made by	Number of varves
Hackensack, N. J.	C. Reeds	6,984
New London, Wis.	E. Antev	6,984
Manitowoc, Wis.	E. Antev	6,942
Menominee, Wis.	E. Antev	6,855
Wrenshall, Minn.	E. Antev	6,700

His time scale shows a longer period than the River Gorge Time Scales. This could be attributed to the clay varves of the same year having been counted at two different locations.

Below the 8,700th varve, approximately, De Geer came upon a giant varve. This evidence fits the theory of the careening globe, because the giant varve marks an abrupt change in the formation of the clay deposits occurring about 8,700 years ago. There is also evidence to indicate that Sweden was located in a different and warmer latitude during the previous epoch of time.

The De Geer Time Scale shows that it is possible to ascertain the approximate duration or length of time of many former epochs, by counting the varves in clay, slate, and shale deposits. Many counts of the number of varves in clay beds have been made and registered, but no systematic attempt has as yet been made to associate the numbers arrived at with epochs of time.

Reappearance of Organic Matter in Clays

THAT vegetation which is blotted out beneath the glaciers of an ice age reappears in the clay beds which have been formed by the silt and other matter carried by summer streams flowing beneath some of the glaciers. This burden of various kinds of matter—carried by flowing waters—is deposited on the bottoms of lakes and rivers when the velocities and carrying capacities of the waters have been lowered or reduced to zero.

Clays are partly derived from vegetation, many of them possessing considerable percentages of organic matter. This appears in colloidal particles—from less than .002 millimeters in diameter down to sub-microscopic. They are the parts of the clays which absorb moisture, and all bear a negative electrical charge when in contact with water, thus showing a certain relationship to cellulose and carbon, both of which are also vegetable matter and acquire negative charges when in contact with water.

The organic matter in clays has been generally attributed to animalculae, algae, and microscopic growths; but tropical forests were another available source of the organic matter found in some of the clay beds developing at the edges of glaciers in former ice ages—such as the five which recently prevailed in northern North America and similar to the ice age now embalming Antarctica and Greenland with glaciers.

Each time the globe has careened, tropical land areas, covered with vegetation and forests, have moved to the polar regions. Theoretically, the vegetation must have become crushed and pulped into colloidal particles by the weight and movement of the overlying ice masses. The lower varves will correspond to the earlier years of the epochs during which they were laid down; the upper layers will similarly correspond to the later years of the same epochs. More organic or vegetable matter will be found in the lower varves and less in the upper varves. Research on the structures of clays may eventually enable us to identify some of the organic matter as having come from crushed forests.

Proofs of the theory that vegetation covered the land when the ice cap first commenced to form should evolve from further study of the varves in the clay beds at Hackensack, N.J., Wrenshall, Minn., and elsewhere.

It is natural to assume that such vegetation, when crushed and ground into colloidal particles, would float in the waters underneath the glaciers, and would thus be carried off and deposited as basic elements of the clay beds. As the limited amount of vegetation was thus disposed of, there must have been less and less available, so that the upper and later varves of the clay beds should show less organic and more mineral silt; the lower or earlier varves, on the other hand, should show more cellulose or organic matter in proportion to the inorganic materials.

Much carbon—generally believed to be of organic origin—is found in the oldest rocks, classed as Pre-Cambrian; but here it is generally found in an amorphous condition, and this may be postulated as having been caused by the pressures and attritions to which the organic materials were subjected. For example, rocks of the Laurentian Shield of Canada are classed as Pre-Cambrian because of the lack of "guide fossils," and for no other major reason. They reveal the scouring of the ice cap of the Ice Age that followed the Life Age which had developed the organic materials ground to a pulp by the ice masses. Carbon appears in some of the black shales of the Lake Superior region. That area was so far inland that the glaciers apparently did not have a chance to purge themselves of the carbon by emptying it into the oceans, as they did to the east, north, and south.

The organic materials of the Laurentian Shield have become so crushed and reduced in the tillage that few organic forms can be identified in those rocks. It is probable that the same kind of pulpification of the tillage is in process, right now, at the bottom of the Antarctic Ice Cap—described later under the section on POLAR REGIONS—and it is questionable whether any guide fossils will be found in the rocks that form the floor of the ice bowl of the Antarctic Ice Cap.

Under the presently accepted theory the rock floor of that ice cap should therefore automatically be classified as Pre-Cambrian, due to lack of guide fossils on its surface areas.

Georges Cuvier of France and William Smith of England announced their independent discoveries, at the close of the eighteenth century and beginning of the nineteenth, to the effect that each stratum of the earth contains fossils peculiar to itself, and that the successive earth layers can be classified accordingly and to some extent dated as to age.

Cuvier found bones of mammoths and of many other extinct prehistoric species of life, and also of many extant species, in the different underground layers of the earth in the environs of Paris. He revealed that a typical series of successively created earth strata shows:

earth layer with fresh-water shells, indicating that a lake had once existed there;
earth layer with marine shells, indicating that the area had once been part of the ocean bed;
earth layer with fresh-water shells;
earth layer of marl;
earth layer with marine shells;
earth layer of clay—no shells;
earth layer of chalk, formed from skeletons of globigerina, which once lived in the ocean.

Cuvier saw with his own eyes and reported the effects of cataclysms in the formations of the layers of the earth. He found that the changes brought about had been sudden, without gradation. He looked for a possible cause and referred to the successive catastrophic changes as revolutions of the earth. He conjectured that the North Pole once had been in the area of the Sandwich Islands (Hawaii).

Since Cuvier's days geophysical discoveries of major importance have been made; they include:

Evidence of Careenings of the Globe 69

The Antarctic Continent	—	1820
Ice Ages	—	ca. 1845
Wobble of the Globe	—	ca. 1885
South Pole Ice Cap	—	Recent
Continuous Growth of Ice Cap	—	New
Continuous Creation of Earth Materials by Photosynthesis	—	New

Much of the mystery previously connected with earth strata, and the problem why successive types of fossils appear therein, are fully explained when these new discoveries are added to those reported by Cuvier.

A communication from the Chief, Paleontology and Stratigraphy Branch, U.S. Geological Survey, states: "The paleontological collections of the U.S. Geological Survey verify that in some localities in the United States and its territories rock strata containing alternating horizons of marine and non-marine fossils do occur."

Again the usefulness of identifying the different species of fossils in each earth stratum is emphasized for horizon* markers, and lists of fossils are given for each formation, by W. M. Winton and W. S. Adkins in *University of Texas Bulletin*, No. 1931, June 1, 1939. They state: "Some fossils appear in recurrent zones, that is, zones between which the fossils in question have never been found."

Atlantis—Plato's legendary land—receives a theoretical validation by the discovery of fresh-water types of diatoms at the bottom of the Atlantic Ocean. Geologists of the Riksmuseum, Stockholm, Sweden, have examined cores taken from the sea bottom of the tropical Atlantic Mid-Ridge, about 12,000 feet below sea level, and have identified algae exclusively of fresh-water origin; this is proof that this area with its fresh-water lake in which the diatoms lived was once above sea level.

---

* Geological term: deposits of certain period, identified by fossils present.

The change in altitude is postulated to have occurred about 7,000 years ago, when the great Sudan Basin Ice Cap, which grew at the North Pole of Figure of the last previous epoch, reached maturity and was moved to the tropics. The fresh-water lake-land—with its diatoms—was at that time rolled around to its present tropical underwater position on the globe. The former position of this lake-land is determined by its distance from the great Sudan Basin of Africa, which, as we have seen, is a telltale depression in the land created by the weight of the North Pole Ice Cap of that epoch.

In theory, this lake-land area was formerly positioned on the globe, in relation to the last previous North Pole of Figure, at about where the State of Oregon is now located in relation to the present North Pole of Figure. It was transposed from about 44° N. Latitude and 120° W. Longitude to where it is now located at about 14° N. Latitude and 30° W. Longitude. It was moved into the bulge of the earth at a latitude where the ocean level is about four miles higher than it was at its previous latitude (four miles further from the center of the earth); thus, quite naturally, it is now under water.

Cores from the ocean bottoms, recently taken in the Arctic and the southeastern Pacific Oceans, have been dated by radium chemical analyses as follows:

ARCTIC OCEAN BOTTOM CORES V. N. Saks			PACIFIC OCEAN BOTTOM CORES Jack L. Hough	
Horizon number	Contain foraminifera (small marine life)	Elapsed time from present (thousands of years)	Horizon number	Elapsed time from present (thousands of years)
1	Yes	9–10	1	11
2	No	9–10, 18–20	2	15
3	Yes	18–20, 28–32	3	26
4	No	28–32, 45	4	37
5	Yes	45–50	5	51
6	No	over 50	6	64

(Saks, Belov and Lapina in *Our Present Concepts of the Geology of the Central Arctic*, translated from Russian, in publication T 196 R, Defence Research Board, Canada; and Jack L. Hough in *Journal of Geology*, May 1953, No. 3.)

The Russian scientists list alternating horizons as cold and warm. Numbers 2–4–6 are listed as cold. Numbers 3 and 5 are listed as warm. This is followed by the assumption that when the climate was warm foraminifera were present and when cold the foraminifera were absent in the core sections. This is obviously an erroneous assumption, because foraminifera are reported present in horizon No. 1 and we know that the climate is now cold.

Commenting on the lowest horizon reached (No. 6), they state "It seems that . . . a considerable part of the Arctic Shelf was dry land." The absence of foraminifera in certain sections of the cores indicates that the areas were not, at that time underwater. Foraminifera are found in both warm and cold waters. A dry climate, with sparse rainfall is indicated for the core sections without foraminifera, because the arctic area is surrounded by continents.

Many scientific papers have contained reports of the different kinds of foraminifera growing in cold and in warm ocean waters. Their presence in the successive strata, found in cross sections of cores taken from the sea bottoms, helps to identify the successive cold and warm sequences of former life at a particular location.

Alternating horizons of the earth's strata with marine and non-marine fossils are not peculiar to arctic regions but are observed in many regions. Drillings, mine shafts, and ravaged cliff sides in many random locations have disclosed marine and non-marine strata in alternating layers, and also alternating cold and warm climate fossils, indicating recurrent relocations of latitude and longitude for all areas of the earth's surface. For example, the ratio of Oxygen 18 to Oxygen 16 in calcium carbonate ($CaCO_3$), globogerinaidae shells, is a function of water temperatures at the time of the growth of the sea shells—a sort of geological thermometer.

A communication from Captain Charles W. Thomas, (now rear admiral retired, U.S. Coast Guard), a noted ice navigator of both the antarctic and arctic regions, states that cores taken from the ocean bottom off the coast of Antarctica and examined by him lead him to conclude that the South Pole Ice Cap is not of great antiquity, but that it is a recent phenomenon, its age being no more than a few thousand years.

The cores showed the ocean bottom to have been formed in layers. The top layer contains cold-water radiolarians and deposition of ice-transported sediments. Below that layer is a layer from which the cold-water diatoms are missing; but they occur again in a lower layer.

The repetitive occurrences, in alternate layers, of approximately the same fossil materials in the earth's strata—disclosed by borings made at different places on the earth's surface—confirms the theory of the successive careenings of the globe. Comparing fossils of fresh-water and salt-water foraminifera, diatoms and algae furnishes clues to the conditions under which each horizon of the strata was formed.

The fossils found in successive earth strata testify to the fact that the layers of earth under any particular land area of the present time have been located during former epochs at many different places relative to the axis of rotation of the earth; the fossils show that these earth strata have been both ocean beds and upland areas in successive epochs, and they also confirm the fact that life existed in tropical, temperate, and cold climates as evidenced by the successive strata.

The fossils testify to the rotation of the earth on successive random Axes of Figure because the variations exhibited by the fossils in the successive layers indicate changes in climate as well as changes from upland to marine locations, and vice versa; and the combination of a change in climate and of a change to or from a marine location can be accounted for only by a change in the location of the Axis of Figure of the earth.

The following tabulation is the driller's record of the deep boring at Spur Ranch, near Rotan, in Fisher County, Texas. It is taken from an article by J. A. Uddin in No. 365 of *The Univer-*

Evidence of Careenings of the Globe

sity of Texas Bulletin, Scientific Series 28, 1914. The drilling was carefully supervised for the purpose of getting a typical picture of the earth conditions underlying a spot selected at random.

	Feet below surface from	to	Thickness
1. Brown soil	0	2	2
2. White porous material	2	6	4
3. Yellow sand	6	16	10
4. Sand and gravel, water	16	23	7
5. Hard concrete of light color	23	27	4
6. Tough red clay	27	53	26
7. Hard concrete	53	65	12
8. Isinglass (selenite) and red clay	65	75	10
9. Hard, flinty rock	75	85	10
10. Red clay and red sand rock	85	98	13
11. White chalky rock	98	101	3
12. Isinglass (selenite)	101	108	7
13. Red clay and red sand rock	108	115	7
14. Isinglass (selenite)	115	119	4
15. Red sand rock, thick streak of red clay	119	135	16
16. Red clay, thin streak of blue clay	135	137	2
17. Red clay and sand rock	137	149	12
18. Red clay and isinglass (selenite)	149	153	4
19. Red sand and clay	153	192	39
20. Isinglass (selenite)	192	199	7
21. Red gumbo	199	221	22
22. Isinglass (selenite) and gypsum	221	223	2
23. Red gumbo	223	239	16
24. Isinglass (selenite)	239	254	15
25. Soft red sand rock	254	272	18
26. Soft red clay	272	285	13
27. White flinty rock and isinglass (selenite)	285	298	13
28. Sand, salt water	298	330	32
29. White flinty rock	330	403	73
30. Red sand rock	403	468	65
31. Hard gray sand, and red sand	468	532	64
32. Soft white clay	532	538	6

	Feet below surface from	to	Thickness
33. White hard flinty rock	538	540	2
34. White tough rock	540	568	28
35. Hard white flinty rock	568	570	2
36. Salt rock	570	580	10
37. Brown sand rock	580	586	6
38. Hard white flinty rock	586	596	10
39. Brown sand rock	596	603	7
40. Tough white rock	603	624	21
41. Hard white flinty rock	624	628	4
42. Hard brown sand rock	628	633	5
43. Salt rock. No sample	633	638	5
44. Light soft rock	638	645	7
45. Hard sand rock	645	674	29
46. Notes wanting	674	688	14
47. Hard sand rock	688	715	27
48. Soft sand rock	715	725	10
49. Soft white rock, hard in streaks	725	732	7
50. Salt rock	732	741	9
51. Hard concrete sand rock	741	773	32
52. White flinty rock	773	778	5
53. Concrete sand rock	778	804	26
54. Sand rock and red gumbo	804	812	8
55. White flinty rock	812	816	4
56. Red sand rock	816	853	37
57. White flinty rock	853	858	5
58. Red sand rock	858	931	73
59. Hard blue rock	931	932	1
60. Notes wanting	932	958	26
61. Red sand rock	958	1113	155
62. Gray lime	1113	1117	4
63. Red sand rock	1117	1123	6
64. Gray lime rock	1123	1125	2
65. Red sand rock	1125	1174	49
66. Soft white rock	1174	1222	48
67. Gray lime rock	1222	1235	13
68. Soft white rock	1235	1250	15
69. Hard gray rock	1250	1252	2

Evidence of Careenings of the Globe 75

	Feet below surface from	to	Thickness
70. Hard limestone	1252	1270	18
71. Very hard lime rock	1270	1272	2
72. Hard limestone	1272	1302	30
73. Very hard limestone	1302	1309	7
74. Hard limestone	1309	1313	4
75. Hard blue rock	1313	1327	14
76. Hard limestone	1327	1335	8
77. Blue rock	1335	1337	2
78. Hard limestone	1337	1341	4
79. Somewhat soft limestone	1341	1347	6
80. Very hard limestone	1347	1349	2
81. Lime and blue rock	1349	1364	15
82. Hard lime rock	1364	1370	6
83. Blue lime rock	1370	1376	6
84. Hard lime rock	1376	1390	14
85. Limestone	1390	1391	1
86. Hard limestone	1391	1397	6
87. Hard limestone with soft blue streaks	1397	1403	6
88. Hard limestone	1403	1419	16
89. Lime rock	1419	1425	6
90. Hard lime rock with soft streaks	1425	1433	8
91. Hard lime rock	1433	1454	21
92. Hard lime rock with soft streaks	1454	1461	7
93. Hard limestone	1461	1478	17
94. Very hard rock	1478	1483	5
95. Hard rock	1483	1502	19
96. Sand, rock fossils	1502	1503	1
97. Blue rock	1503	1506	3
98. Sand, lime, and blue rock	1506	1510	4
99. Hard blue rock	1510	1514	4
100. Blue and gray rock	1514	1520	6
101. Hard gray rock	1520	1523	3
102. Very hard gray rock	1523	1525	2
103. Hard gray rock	1525	1538	13
104. Blue and gray sand rock	1538	1546	8
105. Blue sandy and slaty rock	1546	1551	5
106. Blue sandy rock	1551	1554	3

	Feet below surface from	to	Thickness
107. Hard gray rock	1554	1555	1
108. Gray and blue hard rock	1555	1558	3
109. Hard gray rock	1558	1560	2
110. Hard gray and blue rock	1560	1563	3
111. Very hard gray rock	1563	1575	12
112. Very hard gray flinty rock	1575	1579	4
113. Gray, blue, and yellow rock	1579	1581	2
114. Hard blue rock	1581	1595	14
115. Gray and blue rock	1595	1599	4
116. Blue rock	1599	1600	1
117. Hard gray rock	1600	1619	19
118. Gray and blue rock	1619	1631	12
119. Hard blue rock	1631	1639	8
120. Hard blue and gray rock	1639	1645	6
121. Hard gray rock	1645	1651	6
122. Very hard gray rock	1651	1655	4
123. Hard gray rock	1655	1668	13
124. Blue and gray rock	1668	1676	8
125. Hard blue rock	1676	1688	12
126. Gray and blue rock	1688	1703	15
127. Very hard flinty blue rock	1703	1704	1
128. Very hard sand rock above and then very hard sand and flint rock. Very rough. Rock seemed to have a split in it	1704	1705	1
129. Gray rock. (Mr. W. E. Wrather, who examined this piece of core, describes it as a rough-grained, hard, cemented sand rock).	1705	1707	2
130. Hard blue and gray rock	1707	1730	23
131. Very hard blue flinty rock	1730	1738	8
132. Hard blue rock	1738	1741	3
133. Hard blue and gray rock	1741	1780	39
134. Hard flinty rock	1780	1783	3
135. Hard gray and blue rock	1783	1794	11
136. Hard blue rock	1794	1799	5
137. Hard blue and gray rock	1799	1803	4

Evidence of Careenings of the Globe

	Feet below surface from	to	Thickness
138. Hard gray rock, quit in very hard sand rock	1803	1805	2
139. Very hard sand rock. Had split in it. Very rough.	1805	1806	1
140. Upper six inches very hard sandy, flinty rock, rough, had crack in it. Lower two and a half feet was very hard blue flinty sand rock	1806	1809	3
141. Very hard blue sand rock	1809	1810	1
142. Hard blue rock	1810	1816	6
143. Hard gray and blue rock. Quit in flint at 1823	1816	1823	7
144. Very hard sand and flint rock	1823	1824	1
145. Hard sand and flint	1824	1825	1
146. Blue rock	1825	1826	1
147. Hard flint rock	1826	1827	1
148. Hard sand and flint rock in the upper six inches, then flint sand and blue rock	1827	1830	3
149. Blue rock with flint at bottom	1830	1838	8
150. Flint and blue rock	1838	1845	7
151. Gray and blue rock	1845	1851	6
152. Hard blue rock with streaks of flint	1851	1855	4
153. Gray and blue rock	1855	1860	5
154. Hard gray sand and flint	1860	1862	2
155. Very hard sand and flint and very rough sand and flint	1862	1863	1
156. Flint and sand a few inches, then blue rock	1863	1864	1
157. Blue rock	1864	1874	10
158. Hard blue rock and flint rock	1874	1877	3
159. Blue rock with sand and very hard flint rock at bottom	1877	1884	7
160. Hard blue rock	1884	1898	14
161. Gray and blue rock. Some sand in it	1898	1910	12
162. Blue rock, not very hard	1910	1936	26
163. Hard gray rock	1936	1938	2

	Feet below Surface from	to	Thickness
164. Very hard blue rock	1938	1952	14
165. Flint and blue rock	1952	1955	3
166. Blue rock	1955	1964	9
167. Hard blue rock	1964	1969	5
168. Blue and gray rock	1969	1975	6
169. Hard gray and blue rock; 3 feet gray above 2 feet blue below	1975	1980	5
170. Hard gray and blue rock, gray rock and flint, and sand rock	1980	1988	8
171. Very hard sand and blue rock	1988	1992	4
172. Very hard blue and gray rock	1992	2000	8
173. Grayish blue and gray rock, with flint below	2000	2007	7
174. Very hard flint and sand rock	2007	2008	1
175. Flint and blue rock, very hard	2008	2011	3
176. Very hard blue rock	2011	2014	3
177. Gray and blue rock	2014	2027	13
178. Hard gray rock with streaks of blue	2027	2032	5
179. Hard blue rock with flint in lower part	2032	2036	4
180. Hard blue rock with streaks of flint	2036	2041	5
181. Hard blue rock	2041	2042	1
182. Blue shale	2042	2047	5
183. Soft red sand rock, water	2047	2049	2
184. Blue and gray rock	2049	2050	1
185. Hard gray and blue rock	2050	2059	9
186. Very hard blue rock	2059	2063	4
187. Flint	2063	2064	1
188. Blue and gray rock	2064	2068	4
189. Soft red sand rock, hard in streaks	2068	2107	39
190. Red sand rock and hard gray lime rock	2107	2115	8
191. Very hard gray limestone, almost flint	2115	2126	11
192. Blue rock	2126	2128	2
193. Gray, blue, and red sand rock	2128	2131	3
194. Hard red sand rock	2131	2162	31
195. Red sand rock, not very hard	2162	2176	14
196. Hard red sand rock	2176	2204	28
197. Very hard sand rock	2204	2209	5

Evidence of Careenings of the Globe

	Feet below surface from	to	Thickness
198. Very hard red sand rock	2209	2211	2
199. Hard blue lime and flint rock	2211	2214	3
200. Very hard flint rock (three days' drilling)	2214	2216	2
201. Very hard sand and flint rock (three days)	2216	2219	3
202. Blue limestone	2219	2223	4
203. Very hard flint and limestone	2223	2224	1
204. Very hard limestone	2224	2226	2
205. Very hard blue limestone and flint	2226	2236	10
206. Very hard limestone and flint	2236	2239	3
207. Very hard blue limestone and flint	2239	2240	1
208. Very hard sand and flint rock	2240	2242	2
209. Very hard sand rock	2242	2243	1
210. Very hard sand and flint rock	2243	2244	1
211. Sand and flint rock (core)	2244	2250	6
212. Very hard sandstone (core), much pyrite near this depth reported by Minihan	2250	2270	20
213. Hard blue lime rock (core)	2270	2274	4
214. Blue limestone	2274	2276	2
215. Red sandstone	2276	2278	2
216. Hard lime rock	2278	2281	3
217. Very hard lime rock	2281	2287	6
218. Very hard limestone and flint	2287	2291	4
219. Very hard blue lime rock	2291	2296	5
220. Very hard lime rock	2296	2298	2
221. Very hard lime rock and flint	2298	2300	2
222. Hard lime and flint rock (six days' drilling)	2300	2307	7
223. Very hard limestone and flint rock	2307	2312	5
224. Very hard blue lime rock	2312	2322	10
225. Hard blue lime rock	2322	2329	7
226. Red sand rock	2329	2331	2
227. Hard blue lime rock	2331	2333	2
228. Very hard blue lime rock	2333	2343	10
229. Very hard blue lime rock, almost flint	2343	2348	5

	Feet below surface		Thick-
	from	to	ness
230. Hard limestone	2348	2362	14
231. Hard blue limestone	2362	2381	19
232. Blue limestone	2381	2383	2
233. Hard limestone	2383	2392	9
234. Red sand rock and limestone	2392	2395	3
235. Blue limestone	2395	2396	1
236. Red sandstone and blue limestone	2396	2401	5
237. Blue limestone	2401	2413	12
238. Very hard limestone	2413	2416	3
239. Blue limestone	2416	2429	13
240. Hard limestone	2429	2442	13
241. Blue limestone	2442	2450	8
242. Lime and red sand rock	2450	2466	16
243. Hard blue sand rock	2466	2472	6
244. Blue sandstone and limestone	2472	2480	8
245. Limestone	2480	2487	7
246. Blue limestone	2487	2535	48
247. Red sandstone and limestone	2535	2551	10
248. Limestone	2551	2560	9
249. Blue limestone	2560	2599	39
250. Lime and red sandstone	2599	2612	13
251. Blue limestone	2612	2622	10
252. Lime and blue sandstone	2622	2640	18
253. Blue sand and red sand rock	2640	2653	13
254. Red sand and lime rock	2653	2664	11
255. Soft red sand rock	2664	2673	9
256. Blue limestone	2673	2677	4
257. Blue shale	2677	2682	5
258. Limestone	2682	2685	3
259. Blue sand rock, very hard	2685	2694	9
260. Blue sand rock	2694	2701	7
261. Lime and brown sand rock	2701	2716	15
262. Hard brown sand rock	2716	2735	19
263. Brown sand rock	2735	2744	9
264. Soft gray sand rock, hard streaks	2744	2751	7
265. Brown sand rock, hard	2751	2802	51
266. Brown sand rock	2802	2969	167

		Feet below surface from	to	Thickness
267.	Hard brown sand rock	2969	2975	6
268.	Very hard brown sand rock and flint	2975	2980	5
269.	Anhydrite, water seep	2980	2995	15
270.	Limestone	2995	3045	50
271.	Anhydrite	3045	3046	1
272.	Limestone	3046	3060	14
273.	Hard blue shale with streaks of lime	3060	3075	15
274.	Streaks of anhydrite and hard limestone	3075	3125	50
275.	Limestone, hard	3125	3141	16
276.	Limestone	3141	3180	39
277.	Brown limestone	3180	3185	5
278.	Limestone	3185	3200	15
279.	Limestone and anhydrite	3200	3205	5
280.	Limestone	3205	3210	5
281.	Limestone, very hard	3210	3215	5
282.	Limestone	3215	3240	25
283.	Limestone and anhydrite	3240	3245	5
284.	Limestone	3245	3255	10
285.	Brown limestone	3255	3260	5
286.	Limestone	3260	3280	20
287.	Brown limestone	3280	3290	10
288.	Limestone	3290	3320	30
289.	Limestone	3320	3340	20
290.	Brown limestone	3340	3345	5
291.	Limestone	3345	3350	5
292.	Brown limestone	3350	3355	5
293.	Limestone	3355	3363	8
294.	Very hard brown rock	3363	3371	8
295.	Limestone	3371	3512	141
296.	Very hard limestone	3512	3521	9
297.	Very hard brown limestone	3521	3540	19
298.	Limestone	3540	3667	127
299.	Blue shale	3667	3669	2
300.	Limestone	3669	3752	83
301.	Very flinty limestone	3752	3763	11
302.	Hard limestone	3763	3791	28

	Feet below surface from	to	Thickness
303. Limestone	3791	3842	51
304. Brown and hard limestone	3842	3850	8
305. Very hard limestone	3850	3858	8
306. Limestone	3858	3926	68
307. Hard limestone and some pyrite	3926	3932	6
308. Limestone with a great deal of pyrite	3932	3947	15
309. Very hard limestone and pyrite	3947	3952	5
310. Limestone	3952	3964	12
311. Brown limestone with pyrite	3964	3975	11
312. Limestone	3975	3986	11
313. Limestone with pyrite	3986	3994	8
314. Limestone	3994	4020	26
315. Hard limestone	4020	4045	25
316. Limestone	4045	4075	30
317. Very hard limestone	4075	4076	1
318. Limestone and anhydrite	4076	4088	12
319. Gray limestone	4088	4152	64
320. Very hard limestone	4152	4168	16
321. Limestone	4168	4215	47
322. Hard limestone	4215	4218	3
323. Limestone	4218	4263	45
324. Brown limestone	4263	4278	15
325. Limestone	4278	4288	10
326. Gray limestone	4288	4305	17
327. Limestone	4305	4325	20
328. Very hard limestone	4325	4332	7
329. Hard limestone	4332	4350	18
330. Limestone	4350	4389	39
331. Limestone and shale	4389	4398	8
332. Limestone, streaks, dark shale	4398	4407	9
333. Dark shale and limestone	4407	4431	24
334. Dark shale with streaks of limestone	4431	4470	39
335. Limestone and dark shale	4470	4475	5
336. Limestone	4475	4479	4
337. Limestone and shale	4479	4489	10

These carefully compiled drill data show that there were 337

Evidence of Careenings of the Globe

strata in 4,489 feet—an average of thirteen feet per stratum. Some of the lesser thicknesses—especially where they occur in sequence—may represent time periods of only fractions of epochs. Some of the greater thicknesses may have resulted from drilling through slanting strata. These two different conditions may average out; but it is an assumption that can be corrected when better data become available. On this assumption we will base our estimate of the age of the earth and the age of the oldest rocks that have been sampled.

The nature of the earth conditions underlying a section of the Rocky Mountains, is indicated by the record of the drilling of a water well, furnished by Mr. N. W. Draper and taken from *Colorado Geological Survey, Bulletin 28, 1925*. The well is located 1½ miles south of Grand Junction, in west-central Colorado, just west of the Continental Divide.

Material	*Level*	*Thickness*
Rock and gravel	20– 70	50 feet
Red rock (shale)	70–115	45 feet
Hard white sandstone	125–195	70 feet
Blue shale	195–220	25 feet
Blue shale (lighter color)	220–270	50 feet
Soft white sandstone	270–280	10 feet
Blue shale	280–400	120 feet
Red rock (shale)	400–415	15 feet
Blue shale	415–445	30 feet
Red rock (blood red)	445–465	20 feet
* * * * *		
Red sandstone	815– 870	55 feet
White sandstone	870– 930	60 feet
Red sandstone	930–1190	160 feet
White sandstone	1190–1193	3 feet
White material (like lime)	1193–1197	4 feet
Blood-red material	1197–1210	13 feet
Granite	1210–1213	3 feet

As an example of the earth conditions that lie under a section of the Appalachian Mountains, I reprint here a part of the report of a typical boring, taken from *West Virginia Geological Survey, County Reports*, 1921, Nicholas County:

Materials	Thickness in feet	Total feet
Slate and lime shells	25	905
Lime, hard, gray	25	930
Sand, white, Rosedale salt	120	1,050
Slate & lime shells	65	1,115
Sand	5	1,120
Lime, black	30	1,150
Sand, gray	40	1,190
Slate and lime shells	10	1,200
Red rock	25	1,225
Slate and shells	20	1,245
Lime, gray	50	1,295
Red rock	47	1,342
Slate and lime shells	43	1,385
Red rock	35	1,420

The log of the 20,521-feet-deep well drilling by the Superior Oil Company, in Sublette County, Wyoming—setting a record for depth up to 1950—shows for the last two miles "Alternating sandstones and gray shale with sandy shale and shaley sand to total depth."

The presence of successive repetitive earth strata is indicated by the records of drillings and borings for oil, minerals, and water, and also by mine shafts, in all parts of the world. Practically all the records show that the borings have encountered sedimentary formations in layer after layer.

These records confirm the fact that the globe was built up stratum by stratum, under conditions which were changing constantly for any one area, thus confirming the repetitive careenings of the globe.

Drill logs also disclose that there is an apparent tendency for the globe to repeat its careenings, for a time, over almost the same reel and re-reel, as disclosed by the recurrence of identical materials in its alternate layers.

These facts support the evidence found in Nova Scotia,

referred to above, which contain ten layers of fossil trees—with eleven layers of barren rock between and above and below—and which indicate that the globe careened back and forth within a certain definite pattern or cycle during those epochs.

The records also support our deductions based on the 27 layers of fossil trees in Yellowstone Park, the nineteen layers of coal in Nova Scotia, at the Bay of Fundy, and the successive earth strata with fossil trees reported at frozen Wood Hill in the New Siberian Islands.

Similarly, many coal beds occur one above the other, often with frigid zone materials separating by very sharp cleavage planes quite a number of the strata—and then above and below there are materials which are the accumulations of entirely different conditions of latitude and environment.

Magnetic Rocks

TELLTALE magnetic rocks found in North America and Europe show that in previous epochs, between the recurrent careenings of the globe, they were magnetized in directions different from that in which the earth's electric currents are now magnetizing similar rocks.

Earth electric currents are today magnetizing various types of rocks so that they will point north-south when freely suspended. They are composed of magnetic iron oxide, or magnetite, and have been called natural magnets. They are believed to have been the first compasses used by man.

The angular direction of the magnetic pointings of many of these old rocks are now randomly oriented to the present polar pointings—indicating that in former epochs the North and South Magnetic Poles occupied entirely different positions on the surface of the globe than they do now.

Some of the nonconventionally pointing magnetic rocks are found to be slanted obliquely toward the present ground surfaces, indicating that there have been geological disturbances since they were formed and magnetized in horizontal layers.

Thirteen locations of nonconventionally pointing magnetic rocks have been tabulated by S. K. Runcorn in *Nature Magazine*, September 3, 1955, page 425. He classified rocks of eight geological eras—from Pre-Cambrian to Triassic occurring in Great Britain, North America, and other countries.

John W. Graham, in *Journal of Geophysical Research* of September, 1955, page 327, states that "Enough observations have been made so that there is no longer any question that a useful fraction of old rocks retain to this day the magnetisms they received in remote times."

The most logical explanation for the telltale randomly oriented magnetic rock materials is the recurrent careenings of the globe. The variation in directional pointings of magnetic rocks in old formations is a corollary and proof of the frequent shiftings of the earth's Axis of Figure—caused by the careenings of the globe. Earth electric currents are discussed more fully under "Volcanoes and Hot Springs" (page 236), in Part III— "Origin of the Earth's Materials."

Minerals

CHARACTERISTIC formations—or manner of occurrence in nature— of certain minerals, such as coal, oil, salt, gold, fit naturally and perfectly into the pattern of the theory of the recurrent careenings of the globe. Nothing but such careenings explain the locations and forms of these minerals.

COAL Coal is found in all parts of the world, including the antarctic continent, the arctic islands, Greenland, Alaska, and in all of the temperate and tropical zones. It is being mined under the bed of the Pacific Ocean at Lota, Chile, and under the Atlantic Ocean off England, Nova Scotia, and elsewhere. It is mined in the Rocky Mountains, in the Appalachian Mountains, in the Urals and many other mountains.

The coal beds found in polar regions are the results of vegetable growths which accumulated when those areas enjoyed temperate or tropical climates. Layers of coal were formed from masses of vegetation, consisting of leaves, sticks, and trees, which had become water-logged and then sank to the bottom of depressions in the land—such as swamps, lakes, and rivers—just as vegetable mucks accumulate at similar locations in our own time.

This process resulted in the accumulation during the entire epoch of a bottom layer of vegetable matter, with some mineral contents. All other detached vegetation was exposed to the air and was slowly burned up by oxidation, just as is happening all around us today.

Vegetable muck deposits, which are now found as coal, have been protected from oxidation, i.e. slow combustion, by being covered with superimposed layers of earth materials. The churning-up and dispersal of huge volumes of earth materials by the great deluges which accompany each careening of the globe, create the layered condition now existing in the structure of the earth.

The coal fields of Pennsylvania show five—at some places seven—horizontal layers of coal with layers of shale or slate interleaved between the coal beds. This evidence indicates that these areas were alternately warm or tropical, at which time the vegetable muck which later became coal was gathered, and then polar, when the silt collected, being carried by water and also forming beds at the bottom of depressions.

The vegetable muck was shifted to a polar climate by a careen of the globe and, being covered with water which immediately froze solid, was protected from slow combustion while in the polar region. Into certain of these muck-bottomed depressions there flowed, during the summer, waters containing silt which settled down to form an additional protective covering for the muck. The layer of silt also prevented oxidation when the muck was again careened back to a temperate or tropical zone. The silt eventually became shale or slate.

Having explained the process through which vegetable mucks turn to coal and glacial silts become clay, shale and slate, we notice that there are features common to both formations, and also features peculiar to each. A common feature is that both coal and silts require a depression in the land, because they both form on lake bottoms, etc., from materials which once floated and then sank in these waters.

The features peculiar to each are: (1) coal represents warm climate ingredients which fell into the waters and sank; (2) clay silts represent polar ingredients which were carried by the waters in summer and sank to the bottom, forming a layer above the muck.

Such a series of superimposed alternate layers of coal and slate indicate that the particular area once was a basin of depressed land—created by one of the ancient ice caps once existing in that location.

Thirty coal beds have been found in Pennsylvania and 63 in Nova Scotia in vertical earth strata—indicating possibly as many separate and distant epochs during which these lands were in latitudes and climates suitable to the development and accumulation of the vegetable mucks from which coal is usually formed.

There is always the possibility that a basic coal stratum, laid down in any one epoch, may be found to be divided into seams, separated by interleaved strata and caused by nothing but local disturbances; but there *is* general agreement that each layer of coal buried in the bowels of the earth was once vegetable materials growing upon the earth's surface.

Coal is a product of the land; but oil, salt, and gold appear to be products of the sea. All have been created during successive epochs. The geographical areas in which the greatest supplies are located have been determined by the careenings of the globe. They are all telltale evidence of such careenings.

Starting with the theory of global careenings to explain the perfectly preserved conditions of mammoths, I looked elsewhere in an endeavor to find supporting evidence. Studying the formations containing various minerals I found that coal, oil, salt, and gold stand out as the perfectly natural results of global careenings, but I discovered no other adequate explanation for these materials being located in earth strata at varying depths.

OIL I have come across seven different theories to explain the

Evidence of Careenings of the Globe 89

formation of petroleum—or oil—in the strata of the earth; and only one of them seems to be able to stand the acid test of factual evidence. That theory holds that oil comes from fish. The theory of fish being the origin of oil is ably and adequately expounded in the literature on the subject; but, in all these treatises there has been a "missing link." A cataclysm was required to kill fish in such enormous quantities, and means for preserving them from decay and oxidation immediately after death were also necessary. That link is now adequately supplied by the theory of recurrent global careenings.

The most widely held theory today is that oil principally has been produced by animalculae; but if that theory is correct, we could expect to find oil widely distributed and not concentrated in certain locations. The theory that oil comes mainly from the fish in the seas of ancient times accounts for its being present in certain localities.

Animal life in the sea is estimated to be immensely greater in total numbers than life on the land, for the sea covers approximately 71 per cent of the surface of the globe and supports life for miles below the surface—whereas animal life on land is confined to a single restricted surface. A small part of the great abundance of marine life has been trapped and converted to oil at the time of each recurrent careening of the globe.

Fishes' graveyards, containing their skeletons abound in successive rock strata the world over, in rocks belonging to all of the various systems of formation. The fish skeletons are found in closely packed layers, in an astonishing variety of different sizes and numbers. Estimates have been made to the effect that the beds must have been many thousands of square miles in extent. The Old Red Sandstone found in all parts of the world has been referred to as belonging to the "Age of Fishes," because the remains of whole shoals of fish are found in it almost everywhere.

A hundred years ago Hugh Miller studied the massed fish graveyards in the Old Red Sandstones of Scotland and concluded that the fish must have remained undisturbed in quiet waters following their death. He reached that conclusion without knowing anything about the careenings of the globe, which produce just those conditions. (*The Old Red Sandstones*, published in 1858.)

Massed skeletons of river bullheads have been found in profusion, with their two spines at nearly right angles to the plates of the head, this being a sign that they died of asphyxiation like the mammoth mentioned on page 20. Masses of fish skeletons with fins spread to the full, not relaxed as in a quiet death, are common; and certain individual specimens have been preserved with traces of color on their skin—showing that they were entombed before decomposition of the softer parts had taken place.

Like coal, oils are found in all geological formations, from the earliest to the latest. Oils are derived from a number of sources—from fish, whales, and other animals, from trees, shrubs, and plants. Commercial oils are manufactured from menhaden or silver herring. Fish are the main source of the mineral oils of the earth; this is shown to be so by an outline of the theory of *why* and *how* fish changed into oil.

Consider what happens when a body of water the size of Lake Superior is careened to a polar area and its waters freeze solid; or, if a section of the ocean becomes landlocked in a polar area after a careen of the globe.

The waters immediately become covered with ice. All kinds of fish, by the thousands, are alive and swim about in the water. This is the sequence of events which follows:

(*a*) The body of water becomes hermetically sealed by the ice cover and the oxygen in the air is prevented from being absorbed by the water;
(*b*) The usual percentage of oxygen present in the water is exhausted by the demands of the organic life;
(*c*) The fish all die of suffocation, and they sink to the bottom;
(*d*) The dead fish do not float to the surface, since the lack of oxygen and the coldness of the water prevent the creation of gases of decomposition;
(*e*) The water itself turns to ice—leaving a residue of its mineral salts and silt as a layer covering the dead fish.

Evidence of Careenings of the Globe

When the masses of dead fish, from former epochs of time, were again returned to a warm climate by the next careen of the earth, the depressions in which they rested continued to be bottoms of lakes; they remained filled with water which gathered more silt while the frozen fish were thawing out. The cell structure of the fish had been expanded and disrupted by freezing, just like the cell structure of an apple is disrupted by frigid temperatures. When thawed out, and subjected to pressures by the overlying strata, the hydrocarbons—oil—readily separated out.

At the bottom of this new lake a layer of muck, dirt, and silt gradually formed, and removed the fish remains one layer farther from the surface of the earth. The oils could not float to the surface because they were underneath a covering layer of residual mineral salts and a layer of silt. The oils, therefore, oozed still farther downward through the earth strata above which they had been formed until stopped by a rock obstruction. Today oils are being found in saturated reservoir rocks and sands just above or below geological obstructions by which they have been trapped.

When oils are found below such obstructions and gush on being tapped, or when they ooze from the ground, such oil motion is reasonably explained as due to changes in local, internal earth pressures and centrifugal forces, these being natural consequences of changes in the location of the earth's Axis of Figure.

For a long time many geologists thought that oil could only be found in "domes." These were searched for underground. Then, oil was discovered in buried and forgotten shore lines of underground seas. Geological horizons of ancient sea bottoms are, at sporadic locations, fish graveyards, and oil has been found in abundance in ancient underground coral reefs.

J. J. Newberry has described fossil fish of all sizes found in Pennsylvania and in surrounding states. See *The Paleozoic Fishes of North America*, in U.S. Geological Survey *Monograph* No. 16, 1889.) Naturally, oil has been found in some of the same locations.

The worldwide distribution of underground oil—like coal,

salt, and gold—confirms the theory of the recurrent careenings of the globe and the build-up of its various strata, epoch by epoch.

SALT The great underground deposits of soluble mineral salts, found in many strata under the surface of the earth, can be rationally accounted for by the theory that they are located in what was once the bottom of a depression in the land which was filled with sea water when a flood inundated the land; or their location marks the bottom of a salt sea of long ago which became landlocked because of the earth's careening, its waters having evaporated or frozen to ice—leaving salts as a residue.

The bottoms of salt beds are lens-shaped. Their cross sections are like those of lakes, and this shows that they were lake or sea bottoms at the time the salt residues were accumulated.

In central New York State seven successive salt beds have been discovered, and in the southwest there are thirty or more separate beds of salt, all derived from the evaporation of sea water, as pointed out by Charles M. Riley in *Our Mineral Resources,* page 259. He states that layers of gypsum underlie the rocksalt in mines; he also stresses the fact that gypsum precipitates from sea water after 37 per cent of the water has evaporated, but common salt does not precipitate (crystallize) until over 93 per cent of the sea water has evaporated. The result of this fractional precipitation is that layers of gypsum are laid down before the salt layers.

GOLD Having seen how fish change into oil when a landlocked section of the ocean or a lake turns into ice, and how salt becomes a residue from evaporation or freezing, we will take a close look at the gold that was present in those same sea waters.

Gold—together with most of the common elements of the earth—is found in the oceans. *The Encyclopaedia Britannica* states that gold is present in minute quantities in most rocks and is widely disseminated in igneous rocks, of which, however, it makes up an extremely small percentage. Gold exists every-

Evidence of Careenings of the Globe

where—in rocks, in sands, and gravels. There are important gold fields in every continent. A rich deposit was found in a bed of lignite in Japan and another in the Cambrian coal fields of Wyoming. Most of the gold crystals in ores and rocks are too small to be seen; but in California there are larger crystals, of the cubic system, an inch or more across.

The gold in the ocean was left behind, like the fish and the salt, when the landlocked seas froze or evaporated and thus abandoned the areas in which they had been trapped. Apparently the gold crystallized out just as the salt did, and many of the larger crystals occur in lodes; these have been attributed to the concentration of the residual liquid before its final disappearance. The distorted, rounded forms of many of the larger crystals are due to the pressure and movement of temporary, overlying glacial ice; while the wide diffusion of gold in rocks throughout the world attests to the frequency of the rollarounds of the globe and indicate the great number of locations in which the gold of ocean waters has been left behind as a residue. For example, gold in South Africa is found in "reefs", which have been developed on a rock stratum which was originally horizontal, but is now tilted. Mine shafts that start at the surface outcroppings follow the downward sloping strata, or synclines, to depths of two miles.

The Ocean

OCEAN FLOOR The floor of the ocean may be likened to mountains, hills, ravines, gorges, plains, and river beds. These areas happen to be submerged in our present epoch, but their topography is not essentially different from the land areas now rising above sea level. The fact that these conditions exist is ample proof that there was a time when they were dry land.

There were also times when the world's present land areas were below sea level; in that era originated the marine fossils now in evidence as well as the strata of limestone rocks which were first created in the ocean by the shells of countless shellfish and corals. These rocks provide positive proof that these areas were formerly under water.

The ocean floor and the beaches of former epochs consisted of sedimentary layers of sand, shells, corals, etc.; they were later metamorphosed into sandstones and limestones many of which are found today as rock in the mountains, while countless others form sections of the earth's upper strata.

What was at one time the floor of the ocean may today be a mountain top. Ovid (born in 43 B.C.), in *Metamorphoses*, Book XV, tells of an ancient anchor found on the very summit of a mountain, and of marine shells lying dead far from the ocean.

Explorers have been puzzled by finding sea shells and other specimens of marine life high up in the Rocky Mountains, in the Appalachian, the Andes, the Himalayas, and other mountain ranges. The location of these marine specimens is readily accounted for by the successive Great Deluges of the earth; during each Great Deluge huge quantities of sea shells and marine specimens were churned up as debris, held in suspension by the rushing flood waters, and then widely scattered over the mountains, plains, and valleys comprising the land areas.

Two enigmas, which have long baffled scientists, resolve themselves automatically in the minds of those who accept the theory of the recurrent careenings of the globe. These two riddles have to do with the Land and Water Hemispheres and with the Rifts.

LAND AND WATER HEMISPHERES Our globe may be divided into a Land Hemisphere, containing approximately 46.6 per cent land and the rest water; and a Water Hemisphere, having only approximately 11.6 per cent land. This geographical fact confirms that the Sudan Basin area of Africa was at the North Pole during Epoch No. 1 B.P. It was there that the polar ice cap of that epoch grew to maturity and created the Sudan Basin.

When the eccentric throw of the rotating weight of the Sudan Basin Ice Cap rolled the earth around until the basin reached its present latitude, the earth careened about 80° and ended Epoch No. 1 B.P.

Evidence of Careenings of the Globe

The city of London is located at the approximate center of the Land Hemisphere; it lies approximately halfway between the North Pole and Lake Chad, and nearly on the line of the Ice Cap's travel.

The existence of a Land Hemisphere is explained by the enormous weight and rotating speed and corresponding eccentric centrifugal force of the migrating Sudan Basin Ice Cap. The Ice Cap traveled with a varying rate of speed as it left the Pole of Spin and journeyed toward the tropics; its speed was that of the earth's surface strata for each latitude—plus the additional speed generated by the careening motion of the globe's surface.

Neither the speed of careening nor the speed of the earth's rotation affected the Sudan Basin Ice Cap when it was at the North Pole. Then it moved only at the slow speed of the wobbling motion of the earth and at the speed created by the distance the Axis of Figure was off center from the Axis of Spin. The speed of careening soon became excessively great, but decreased to practically zero when the Ice Cap reached about 10° Northern latitude, while its speed of motion—due to the rotation of the globe—became about 40 per cent faster than the speed of sound, since it would then move at approximately the speed of the earth's surface at the Equator.

The eccentric centrifugal force created by the motion of the Ice Cap at these great speeds caused the elevation of all the lands of which it was a part. It pulled them upward and outward from the center of the earth, against the force of gravity. At the same time most of the land areas of the opposite hemisphere became submerged in the oceans, and became a Water Hemisphere.

More than three quarters of the land surface of the globe is north of the Equator. This is so because the three most recent ice caps that have caused the globe to careen have been located at the North Pole. It is therefore to be no more than expected that the northern land areas are found to be elevated. The present arrangement of the land masses is evidence of the upward and outward throws of the eccentric centrifugal force of the migrating ice caps of the past.

Rifts

THE UPWARD and outward pulls as well as stresses of the eccentric centrifugal force of the Sudan Basin Ice Cap created during the ice cap's migration put certain tensions in the adjacent earth masses. The great African Rift remains as mute telltale evidence of this. The rock formations thus torn apart tell of the transient forces created by the Ice Cap's migration. It has long been known that tremendous force and tension were required to form the crevasses in the rocks, and that force is now clearly identified. We also identify this rifted area as a section of the globe that was moved from the North Pole to the tropics; it was at the same time moved about thirteen miles further from the center of the earth. The radius of the earth being about 4,000 miles, the area had to be stretched out about 13/4000 squared, more than it had been, for surfaces of spheres are to each other as the squares of their diameters.

The Great African Rift lies to the east of and parallel to the direction of the Sudan Ice Cap's line of travel. It extends both north of and south of the Sudan Basin, ranging from Syria to south-central Africa, a distance of over 4,000 miles. Bailey Willis, in Carnegie Institution of Washington Publication No. 470 (1936), has collected many photographs of rift valleys. In many places the sides are vertical, or nearly so, and are so bare and sharply cut as to indicate the rift's recent geological creation. Some of the rift valleys have normal escarpments.

J. W. Gregory delineates and describes sections of the Great African Rift Valley in his book *The Rift Valley and Geology of East Africa*. Both of these authorities refer to other rifts known in many parts of the world. They describe the Rift Valley as being in some places, a single chasm and sometimes being as wide as the Red Sea; in other places it has been broken into a long, wide chain with numerous chasms. The Rift Valley branches eastward to the mouth of the Gulf of Aden and westward beyond Lake Tanganyika in the rift valley of the central Congo region. The Red Sea is not in a valley; no important rivers

Evidence of Careenings of the Globe

flow into it. It appears to be a crack, approximately 1,250 miles long, in the upper rock surface of the earth, where the earth opened up, due to transient tensions, and stayed open.

The Dead Sea, near the northern end of the Great Rift, is 1,300 feet deep and its surface is 1,293 feet below sea level—the lowest land surface on the earth. The Dead Sea and the River Jordan lie in a narrow valley, so straight and deep that it has been described as a crevasse in the earth's so-called crust. Similarly, Lake Tanganyika, lying in what is one of the southern extensions of the Rift, is 4,190 feet deep and its bottom is 1,664 feet below sea level. The Great Rift is indeed no local fracture, its length being one-sixth of the circumference of the earth.

There is much evidence to show that the rocks in the area have been pulled apart. In many places parallel faults, extending north–south, are arranged like a grid, or like parallel fingers of long, thin rock slices on end, separated by bays of alluvium.

If we look for evidences of earth tensions on the side opposite the Rift, caused by the Sudan Basin Ice Cap reeling southward and causing the land surfaces to stretch, we notice both the main fjords at Oslo running north–south and the English Channel. In the Channel area we find that a great deposit of chalk has apparently been split approximately down the middle, with one half in the cliffs of Normandy facing the other half in the cliffs of Dover.

In view of the above facts, indicating that rifts were created by the tensions resulting from the eccentric centrifugal forces of rotation of polar ice caps in transition, we can confidently look for other rifts caused by former polar ice caps, and we will find them as fjords, chasms, and steep, walled valleys all over the surface of the earth.

Polar Regions

EARLY explorers arrived in Antarctica with comparatively open minds but did expect to find evidence to support the then current belief that the ice mass was the waning remnant of a prehistoric ice age. They discovered physical phenomena which they erroneously concluded were proofs of the ice mass having been larger in former times.

Near the coasts they observed glacially transported boulders perched 1,000 to 1,500 feet above the flowing glaciers, and high up on the sides of mountains they saw the scouring marks, striations made by moving glaciers. They assumed that the Ice Cap must have been at least 1,000 feet higher in former times, and their erroneous conclusion that the Ice Cap was waning became current popular belief. It is possible that they observed the telltale markings of an earlier ice age for Antarctica; as *The Encyclopaedia Britannica* states, "Raised beaches show an emergence of land in Quarternary times and there is evidence of a recent glacial period when the ice sheet on the Palmer Peninsula was 1,000 feet higher than it is now." (Vol. 2, 1959, page 14).

A more recent view holds that these phenomena more probably have been caused by the natural workings of isostacy, the inland ice pressures having caused the extrusion of coastal mountains. The Ice Cap, fed by copious snowfall and almost continuous fall of hoarfrost, appears to be constantly growing in robust health, and not waning. Glaciers, past or present, never could pile up any higher on the sea coasts than they do at present, because any increase in weight makes them flow away faster into the oceans, and the ice is constantly flowing off the land and into the sea. Therefore, glaciers could not have deposited the 1,500-foot-high boulders on the coast nor have caused the scouring marks on the high coastal mountains.

Some of the coastal rocks of yesterday have apparently become the coastal mountains of today. They were spewed up to relieve the tremendous pressures created on the rock floor of the inland ice bowl. Greenland's ring of coastal mountains and its depressed center appears to be a similar example of the workings of isostacy.

Such rock upthrusts receive confirmation from the carcass of a Weddell seal, found by Captain Scott high up in the twin Ferrar Glacier, near the Ross Sea Coast. For a seal to have climbed so high is, of course, absolutely impossible. For a seal to have been reposing on rocks which were flung up to a higher

Evidence of Careenings of the Globe 99

altitude can be explained in the same way as the finding of glacial boulders perched high in the mountains.

Solid rock flows under pressure, but it cannot move downward. Antarctica's interior ice pressures are relieved at the coasts, the rock being burst up into the coastal mountains. Thus added to, these mountains serve to enlarge the area of the continent, and at the same time, they block off or dam the flow of ice. This increases the volume and weight of cold storage ice, which, in turn, acts to produce new coastal mountains. *This is the vicious circle of Antarctica's growth!*

"Ice mountains" and "ice volcanoes" illustrate the vicious circle of the continent's growth. An ice mountain—discovered along the Queen Maud Land coast by the U.S. Navy's 1946-47 expedition—is described by Admiral Byrd as "luminous blue, towering more than two miles high and extending 100 miles along this coast." Others have described it as rising sharply from the ocean depth. It now blocks off the flow of the ice to the sea, increasing the weight of the Ice Cap.

"Luminous blue" signifies deep glacial ice. Placed in a glass of water, it gives off air bubbles as the ice melts, the effervescence being due to the air having been under pressure. Deep glacial ice, now high up in a mountain, can only be accounted for by the theory of underground rock movements and coastal upthrusts.

Ice volcanoes, or "ice bowls"—which pockmark a large area of the Bellingshausen Sea coast—are caused by sudden violent rock flows resulting in pillars, or guyots, that have been thrust up. They block off the flow of ice to the sea, and thus increase the Ice Cap's weight.

The ice appears to have been thrust upward with such speed that the momentum caused the upper sections of the glacial ice to become detached from the parent ice on the floors of the bowls. Some of these upper layers of ice were extruded so violently that they broke into great blocks, the size of houses and ships, some of which landed on the lower ice shelf and some back in the craters.

The rocks under these bowls have been thrust up as the

result of the same processes, described on page 101, that have formed the underground pillars of salt, the clay pipes of the diamond mines, and the ocean guyots—as well as the ice mountains. The contours of the tops of these rock extrusions are either flat or irregular, depending on whether the ground levels from which they were extruded were even or very rough. It is predicted that when some of the sheer wall-like coastal mountains are examined more closely the rock surfaces will be found to look less aged—less eroded and spalled by frost action—than mountains rising above the ice sheet in the inland areas. This will confirm their more recent creation.

Where a long range of mountains—now located inland from the sea coast—lies parallel to it, the presumption is that the inland range at one time was actually at the coast. The mountain range of Queen Maud Land, 100 miles or more inland, is an example.

Greenland's topography indicates that the mountains along its sea coasts have been thrust up by the same process. Its central plateau of ice is about 10,000 feet above sea level and is contained by these mountains, now forming most of its sea coasts and shore lines. In the central areas the rock floor has been depressed below sea level by the weight of the ice.

Antarctica's highest elevation is reported to be approximately 14,000 feet above sea level. Recent depth-recording echo soundings have disclosed that the rock floor, in some central locations, is a mile below sea level. Thus, the maximum ice column may be estimated to be approximately 19,000 feet in height. The resulting pressure on the rock floor is over 7,500 pounds per square inch—over 1,000,000 pounds (500 tons) per square foot at those particular locations, assuming that the ice weighs uniformly 57.5 pounds per cubic foot. To repeat, this bottom pressure appears to find relief at the sea coasts by pushing up the coastal mountains.

It has been noticed that striated glacial markings are found on the rocks forming one side of a glacier-filled valley, and that no markings appear on the softer rocks forming the opposite wall of the same valley. The markings on one side could never have been made by the valley glaciers; if the glaciers had reached higher previously, both sides of that valley would have striated walls. It is more reasonable, then, to assume that the rocks have been forced upward than that the glaciers have sunk down. Apparently the rocks on both sides of the valley have been thrust upward by underground pressure, caused by the central ice weight, the striated one by tilting and the softer, unstriated rocks by direct levitation. It is also postulated that enormous striated rock masses have been carried seaward because of the inland ice pressure.

Starving glaciers have been reported by explorers who have observed cirques and bare mountain walls. But available photographs show that bare walls always occur on the north side, except where flowing glaciers are carrying the ice seaward by the force of gravity. The glacial ice does not appear to be starving on the south side of valleys, where inland ice pressures apparently fill all available space with ice.

Oases, which are limited areas of bare rock and sand, free from snow and ice, are found in the midst of ice-covered areas. Some explorers and writers maintain that these oases indicate that the entire Ice Cap is waning. The phenomena are explained more rationally, however, by the theory of earth electric currents, which heat the land and cause snow and ice to melt. They are the cause of the fumarole of Mount Erebus. (See "Volcanoes and Hot Springs," Part III, page 236.)

Iceland, as an analogy, has both glaciers and hot springs. Many buildings in the city of Reykjavik are heated in winter by the hot water piped from the hot springs. Ice-free areas and frozen lakes are attributed to the same cause that creates the hot springs. It is obvious that the cold-storage facilities of the great antarctic continent are reduced by only a tiny fraction by the relatively small areas of localized heat.

The Creation of Mountains

THE forces of nature react on each other in various ways during the active periods of global careenings. Because of the curvature of the globe the centrifugal forces of the rotating ice caps which initiate the careens soon reach a maximum and then diminish.

When the ice caps have migrated 45 degrees of latitude their centrifugal force responds to the combined motions of careening and rotation. Between the sun latitudes of 45 and 0 degrees they change from being upsetting to being stabilizing forces.

Equatorial bulges then start to form, and the centrifugal forces of the ice caps and of the new bulges of the earth are soon working in unison to bring the reeling motions of the globe to a rapid slow-down and stop.

In the meantime, kinetic energy which has developed in the continental land masses because of their weights and velocities, collides with the combined energy of the newly-generated bulges of the earth and of the ice caps.

The result of these collisions of forces is that the energy of the moving continents is absorbed by the crushing, elevating, and wrinkling of large land areas whose rock strata are crumpled and bent in ridges at right angles to the forces being dissipated.

A striking illustration of the formation of mountain ranges, by the dissipation of the mechanical energy of the rolling earth masses when brought to a halt by a superior force of global stabilization, is the great chain of mountains lying approximately at right angles to the directions of motion of the last three rollarounds of the globe. They extend along a nearly perfect meridian circle skirting the basin of the Pacific Ocean, traversing the west coasts of South America and North America and parts of Asia. The directions in which the three last North Pole ice caps moved, while rolling the globe sideways to its normal direction of rotation, were, (1) On a line from the Hudson Bay Basin of Canada to the Caspian Sea Depression of Russia (which previously had been at the North Pole), (2) From Hudson Bay Basin (which had been at the North Pole) to the Sudan Basin of Africa, (3) From Sudan Basin to present North Pole.

It is customary to refer to Africa as "a plateau of continental size, because its margins are abrupt . . ." (*Elements of Geography*, by Vernon C. Finch and others). The elevation of the continent of Africa is due to the centrifugal force of the most

Evidence of Careenings of the Globe 103

recent roll-around of the globe, which was spearheaded by Africa and its ice cap; as mentioned before, the depression of the great Sudan Basin remains as a telltale evidence of this ice cap.

The wrinkles and folds in the rocks of the earth's upper layers are not uniformly distributed. They are not, for instance, like the surfaces of desiccated plums (prunes) and dried apples, whose wrinkled surfaces do not have fault foldings, as do some of our mountains. Such evidence provided by the earth itself, helps to refute the older theory that the *wrinkling of rocks into mountains* indicates that the earth is shrinking in volume. We now know that it is growing in volume.

Cross sections of the Appalachian mountains and the Alps show shallow folds of rock strata, the result of great compression forces acting tangentially at that point of the earth. Below these folded strata are other earth layers making up entirely different formations; that they are unaffected by the local surface forces is shown by countless illustrations in geological literature. This, in turn, indicates that the folds have been caused by surface forces which did not affect the lower earth strata.

The warping and the folds of localized earth layers become corollary evidence supporting the theory of recurrent careenings of the earth and of the tangentially imposed compression forces. The energy involved in careening earth masses, when brought to a halt, is sufficient to create the observed folds and warpings and is also great enough to change the horizontal position of many earth strata to a vertical one, and also to cause the synclines and the anticlines in the earth's rock structures.

All changes in the surface of the earth are due to forces which even in our day are either active or latent, and these forces will continue to produce similar changes for the duration of the earth's existence.

Mountains of the Rock of Gibraltar type indicate clearly that a vertical geological break or fault has occurred in such rock strata, and that it was caused by an irresistible force; it brought about an upheaval or elevation of the tiered underground earth strata on one side of the fault, exposing the laminated strata in one sheer face or precipice. The other sides of such mountains will usually slope down gradually, the earth's surface being practically as it was before the upheaval, but it is now slanted to the horizontal.

The precipitous face of this type of mountain weathers and erodes increasingly toward its summit, and the area around its base is filled with talus, breccia, and debris—deposited on what was formerly the surface of the earth on the opposite side of the geological fault.

Mountains of this kind are produced not only when the earth careens and then rapidly ceases its reeling motion, but also during ice ages—such as the ones now prevailing in Antarctica and Greenland. The normal workings of isostacy equalize the ice and rock pressures by underground earth movements, including rock fracture and rock flow.

The Ice Cap in Antarctica, for example, exerts a pressure, at sea level, of about 500 tons per square foot on the underground rock materials. This pressure can be relieved only by lateral rock motion in the center and by lateral and upward motions of the rocks at the edges of the continent.

Some parts of the mountains forced upward by the lateral underground flow of rock materials—as occurs in the coastal mountainous areas of Greenland and Antarctica—will be found to lack well-marked stratification, for the stratifications become distorted in those parts of the rock materials which have been forced to flow.

Subterranean pressures have raised up mountains of various shapes. Some, like Gibraltar, appear with slanting elevations placed almost on end; some look like humped mounds; others resemble a long finger of uplifted stratified rock layers, their centers welling up like heads of well-developed cabbage which have been tremendously elongated, and which have divided numerous layers of the upper rock strata.

Sheep Mountain—on Wyoming's Big Horn River—is an example of the last mentioned type. In the course of time the center section, where broken, has eroded away to form a hogback mountain. The harder or more durable rocks of the cracked outer strata have eroded and become ridges paralleling the contour of the main formation.

The unusual appearance of Steamboat Rock at Dinosaur National Monument, Utah, its shape and flat stratifications, resembles a multistoried New York City skyscraper—with setbacks on one face and a sheer flat precipice face on the other. It appears to have been extruded by subterranean pressures seeking relief.

The birth of a different class of small mountains, created by volcanic heat, is described in Part III, in the chapter entitled "Theory of Volcanoes."

The polar ice caps, which have existed in all geological epochs, are a primary instigating cause of the creation of mountains. They produce mountains by the pressures they exert while growing, and by causing the globe to careen when they reach their maturity.

The successive layers of earth strata are clearly seen in many mountains. This indicates that the strata now forced up to the surface were once below the surface. The flat tops of some mountains may have been the original level land surface, or the flat top may be all that was left when the topmost layer or layers of the mountain were carried away by a flood during one of the careenings of the globe, or were floated away on the ice during an ice age.

The Ice Cap in Antarctica is suspected of separating mountain tops from their bases by creating cleavages along the strata planes of the mountain layers. A limestone layer will slowly dissolve in fresh water, which is slightly acid. Such a layer might become so slippery that the side pressures of the ice could cause the whole top section of the mountain to slide sideways and be carried by the ice to the sea coast like any other oversize boulder.

Submerged flat-topped, conical mountains have been found in the ocean. They are called "drowned ocean islands," or "guyots." More than 500 have been charted in the Pacific Ocean during the past decade, and a few in the Atlantic.

The theory that these mountains were extruded and forced

upward either during ice ages or were decapitated by ice during an ice age fits the facts more snugly than any theory yet advanced to account for them. Their flat tops were probably the flat ground level where the extrusion occurred. The "ice bowls," described in the section on Antarctica (page 99) are postulated to have resulted from just such extrusions.

Other well-known types of earth formations fit into the theory like the pieces of a jigsaw puzzle when they are explained as having been caused by the tremendous pressures of overlying ice caps; and no other adequate explanation has previously been advanced to explain the phenomena.

The clay columns or "pipes" in which diamonds are found in Africa and Brazil,—identified by me as beds which underlay ice caps—are composed of silt in nonsolidified but articulate form. The great pressures created when ice caps develop above clay beds cause the particles to rearrange their formation and to be squeezed into and penetrate the upper adjacent materials of the earth, and therefore the clays are forced into the shapes of long pipes.

The particles composing some African pipes have been termed breccia because they consist of many small rock particles of widely varying chemical composition. The fact that the pipes are like cornucopias, with the large end up, is an indication of a sudden upthrust of material under a great pressure which caused the upper part to expand more than the lower part, the reason for this being that the resistance of the lateral rock pressures were less great toward the surface than deep down. The upper surfaces of rock strata in the South African diamond area are marked by sharply defined straight furrows characteristic of the ice ages.

Just why diamonds are found in these clay pipes is quite another subject; but in view of the pressures created by a tremendously heavy overlying ice mass, it seems reasonable to assume that the carbon contents of animal or vegetable items—trapped snugly in the clays—changed gradually into the diamond form of carbon crystallization.

Ocean waters, trapped in landlocked basins following any

one of the many former roll-arounds of the earth, evaporated and left behind vast beds of salt—with flat surfaces and bottoms having lake-like contours. The surfaces became overlaid by some of the many minerals left behind when the sea waters evaporated. These, in turn, were covered by various earth materials during the following epochs between and during the recurrent careens of the globe.

In drilling for oil, sulphur, and other materials, it has been found that pressures—heretofore unidentified—have forced some of the salt upward into straight, smooth, sheer underground pillars. There are usually several layers of the normal rock strata which cap these salt plugs and appear to have spearheaded the advance of the softer pillars as they emerged from the main salt beds, under pressure of the overlying ice.

This release of pressure is somewhat analogous to the popping of the cork from a bottle of champagne. "Ice bowls" and "ice mountains" are similar extrusions forced by the ice cap pressures.

Salt pillars have been found only whenever salt beds have been compressed into dome shapes, and both domes and pillars are natural phenomena explained by ice cap pressures. Domes as well as pillars record the fact that an ice cap once existed in that locality.

Ocean Depths and Mountain Heights

CHANGES in the elevations and depressions on the earth's surface, caused by the locations of the earth's bulge and axis being shifted following each careen of the globe, are limited to 13 miles vertically. The diameter of the earth at the bulge of the Equator is about 26 miles greater than the length of the polar axis, and therefore the maximum change of land elevations and ocean depressions is limited to half of the change in diameter at any given point on the surface of the earth. The total changes on both sides of the globe may add up to about 26 miles, but the changes on any one side will not exceed 13 miles at the points of maximum change.

These changes in elevations and depressions have occurred repeatedly each time that the Axis of Figure has been shifted to a new random location, and today the surface of the earth consists of innumerable elevations and depressions.

The highest mountain peaks are nearly 6 miles above sea level, and the greatest ocean depth almost 7 miles below sea level. They add up to just under 13 miles—just within the permissible theoretical limit.

The Sudan Basin land area, which was at the North Pole during Epoch No. 1 B.P., was moved approximately 13 miles further away from the center of the earth when the globe last careened, while the land areas now at the North and South Poles—which were at that time near the edge of the tropical zone—were moved closer to the center of the earth by about 13 miles.

These were cataclysmic disturbances in the surface layers of the earth; yet, when the various forces involved readjusted the earth's surface in order to eliminate the conflicting pressures of the centrifugal force of rotation, the kinetic energy of motion, and the force of gravity, the total vertical distance between the greatest depressions in the sea bottoms and the highest elevations of mountains remained within the total possible range of 13 miles.

This balancing of existing pressures, effected by rapid movements of rocks and other earth materials, resulted in the first post-flood isostasy or equilibrium of the earth's stratifications. Since then, isostasy has been maintained by the earthquakes which occur every day.

The theory of the recurrent careenings of the globe—caused by the eccentric centrifugal forces of great rotating ice caps—accounts for the limitations found to exist for ocean depths and land elevations. The tremendous forces which have torn earth layers apart and caused mountains to be formed in chains, or ranges, can now be identified. The reason that the floor of the ocean resembles the contours of the land is no longer a mystery. The great ice caps of the recurrent ice ages aid in accounting for many of the geological faults, and for the Ice

Evidence of Careenings of the Globe

Front, which fringes much of Antarctica far out into the oceans. The actual coastline is reported to be mostly unrecognizable, due to the continuity of the ice which extends out into the sea from one hundred to several hundred miles.

The earth has been rotating on its present Axis of Figure for about 7,000 years, as shown by the time scale of Niagara Falls (see the chapter on "Rivers," page 35). The Ice Cap has grown during that period of time to a height of 14,000 feet above sea level. The weight of the South Pole Ice Cap now approximates the astronomical figure of nineteen quadrillion tons—19 followed by 15 ciphers. This figure is derived from the statement of U.S. Coast and Geodetic Survey to the effect that if the ice of Antarctica were uniformly distributed around the earth, it would make a layer 120 feet deep. One arrives at the same figure by assuming a cone two miles high with a base diameter of 2,800 miles.

The heat of the sun—strange as it may sound—causes the great South Pole Ice Cap to grow continuously. If one keeps the globe in one's mind's eye and pictures air currents rising everywhere throughout the temperate and tropical zones—caused by air becoming heated by the sun's rays—one will notice that heated air rises, since it expands and thus becomes lighter. Heavier, colder air flows in beneath it.

In the southern hemisphere, below the Equator, the rising air currents flow south. They cannot flow north, because similar air currents are rising on the north side of the Equator.

Since the earth is a sphere, all of the south-flowing air currents must converge at the South Pole. These air currents meet "head-on" from every direction, and, because of the speeds of their motions, they build up air pressures over the polar areas. These pressures are relieved when the air currents curve downward because of the fact that colder air is heavier; they thus reverse their directions of flow and pour northward at low levels and at very high velocities.

Clearly then, it is the heat of the sun's rays, in other parts of the world, that has caused the great South Pole Ice Cap to grow to be approximately two miles high in approximately 7,000 years. It is also evident that the growth of the Ice Cap started and continues because of the physical property of air to act like a moving conveyor, to absorb water like a sponge when warm, and to wring out the water when cold. The south-flowing air currents become so cold, as they approach the South Pole, that the moisture is wrung out of them.

At Little America the general flow of the air currents is from the north at high altitudes, and from the south at low altitudes. The south-flowing air currents are found at altitudes of 3¾ to 7 miles (U.S. Navy, Hydrographic Office, *Bulletin* No. 138, "Sailing Directions for Antarctica, 1943," page 38). Theoretically, there should be more precipitation of moisture on the inland area than at Little America because the air flowing south is warm, moist, and high up, while the air returning northward is cold, dry, and at a low altitude.

On the journey southward the air becomes increasingly chilled, and naturally forces precipitation. On the return journey northward, the air currents begin to warm and this naturally increases their capacity to retain moisture. Such winds would be expected normally to produce cloudless skies over Little America. The air at Little America is reported very dry although the relative humidity is high, the air holding ⅕ of 1 per cent of moisture. The same air in tropical regions often contains up to 3 per cent of moisture at sea level, or about 15 times as much.

Less precipitation will fall on areas facing the open ocean, as does Little America, than on those facing the continents, of Australia, South America and Africa, remembering that the winds which bring in the moisture and snow are strongest opposite the continents. The continuous fall of snow and hoarfrost produce cold storage ice. A small amount melts in summer, but practically none melts during the winter months when the sun is generally below the horizon.

Studies made during the International Geophysical Year (1957–8) reveal a peripheral zone of maximum collection of water, or "rainfall", which increases inland from the coasts and then decreases at the Pole. Reports of accumulation of water

Evidence of Careenings of the Globe 111

range from 2 inches to 2½ inches at the pole, each year, to 33½ in the area facing Australia, where the great cyclone winds bring in the most moisture.

Today, many scientists find that the ice cap is growing, and several have issued statements regarding the yearly rate of growth. Ten years ago textbooks generally maintained and students believed that it was waning. Our own National Science Foundation prefers to defer any final estimate or statement. They wish to thoroughly analyse the detailed reports of yearly precipitation of moisture and to check it against the ratio of ablation and flow-off of icebergs in many areas. They know that a statement to the effect that the ice cap is growing will be the signal for an all-out attack to halt its growth, at a cost equal to that of a war. If and when they announce that it is growing and is not waning, the rate of its growth will not be as important as the number of years before it, in combination with the wobble of the earth, will cause a roll-around of the globe, with a catastrophic flooding of ocean waters over the land areas.

Just as a growing tree sheds leaves, but a dead one does not, a growing glacier sheds icebergs, but a waning glacier does not. We have records—dating back several hundred years—which show that the South Pole Ice Cap has been shedding great icebergs for as long as men have navigated the adjacent waters. It continues to be a prolific breeder of icebergs, which adds proof to the theory that it is growing and is not waning.

The flow-off of icebergs shows us that the South Pole Ice Cap is bursting at the edges continuously. The extrusion of the ice shelves and the ice cliffs, into the sea, results from the pressures of the inland ice; and the ice pressures result from the weight of the snow and ice continuously building up in the central areas.

A part of the Ross Ice Shelf—containing a section of Admiral Byrd's Little America—has already been extruded so far out into the sea that it has broken off and floated away. The rest of Little America will follow. The shelf ice on which the German Weddell Sea Expedition of 1911–12 was based, and the deck ice used by the Norwegian-British-Swedish Expedition of 1949–

52, have both disappeared by breaking loose, calving, and floating out to sea as icebergs.

An iceberg 208 miles long, 60 miles wide, and extending below the surface about 700 feet, was sighted by the American icebreaker *Glacier* in November 1956. Its area was about that of the states of Connecticut and New Jersey combined. It was a "calf" of the Ice Front. Another iceberg—240 feet above the surface, which indicates probably around 2,000 feet below—was reported by Captain Scott.

A growth of about two feet a year is shown for the Ice Shelf on which Little America is located. In 1929, Admiral Byrd erected two 70-feet-long steel radio towers, projecting 60 feet upward. In 1934 they had been so covered by snow that they projected only 30 feet. In 1947 they were 18 feet high. In 1955, one extended 8 feet, and the other 10 feet, above the level of the Ice Shelf.

A growth rate of about a half foot per year is disclosed by photographs of another part of the fractured front of the high ice cliffs fringing much of Antarctica. Along the coast an average of 80 feet projects above the ocean, on which it floats, and about 160 varves or annual layers of the ice accumulation are visible—indicating about 6 inches per layer. This Ice Front is about 800 feet thick. It is an extrusion of the Ice Cap which extends more than 100 miles out into the sea, but is still attached to the inland ice.

Sir Douglas Mawson has reported his observations of the winds at Adélie Land, which is adjacent to Wilkes Land. In 1911–14 he found that the winds continually poured off the Ice Cap, exceeding a velocity of 90 miles per hour for periods of more than 24 hours, and reaching puffs of 180 miles per hour. The rate averaged 51 miles per hour for the entire year.

These terrific winds at Adélie Land are caused by the nearness of the continent of Australia. This enormous rush of icy, dehumidified air, pouring northward from the Polar Plateau, is the natural return circulating air current, at sea level, of the moisture-laden, heated air which rose over Australia and poured south at high elevations.

Evidence of Careenings of the Globe 113

Antarctica is surrounded on three sides by continents from which heated air currents rise, flow south, converge at the South Pole, turn downward and reverse their directions of flow—after surrendering most of their moisture content to promote the further growth of the great Ice Cap. This makes it appear that the moisture being carried Pole-ward will have become precipitated as snow prior to reaching the Pole. The lower altitude of the South Pole, compared to the surrounding areas at higher latitudes may be due to less precipitation at the polar center.

The Ice Cap Plateau is found to have an apparently sunken surface at the South Pole. A vortex is suggested because the Pole is the center of rotation, and higher elevations, producing greater pressures, occur nearer to the center of the continent. The highest altitude reached by both Scott and Amundsen was at 88° 31'; this spot is at least 1,000 feet higher than the South Pole—which is 9,200 feet above sea level. The sunken polar center of the Ice Cap tends to confirm both a minimum local snow fall and the underground flow of continental rocks. The coastal mountains have been thrust up as a result of the ice pressures on the rock floor of the Ice Bowl. When the central rocks went down the coastal rocks went up.

The base of the great ice pyramid grows ever larger in response to the overlying ice pressures, this being the cause of the lateral underground flow of rock materials. The widening of the base permits the Ice Cap to become higher at the center, while the gradient or slope of the glacial ice remains the same.

The extension of the base allows the weight to become greater; the greater weight, in turn, increases the depth of the dent in the earth's surface, or the Ice Bowl, and this results in an additional flow of underground materials to the coastal areas. What is going on therefore is a progressive broadening of the base, by the widening of the Ice Cap Bowl, with a concurrent increase in the total height of the Ice Cap—a vicious circle of continuous growth.

The tremendous weight of this great Ice Cap is now producing pressures which will result in bulges or adjustments of adjacent earth materials.

The ice-embalmed continent of Antarctica is undergoing slow geological changes, the result of which is a general increase in its area—due mainly to isostasy, or the adjustments of earth materials to natural physical forces. Antarctica thus holds a key position in the impending tragedy—the next great deluge of the earth.

Analogous are the Hudson Bay Basin in Canada and the Sudan Basin in Africa. The heights of the perimeter of the Laurentian Shield which encompass Hudson Bay and bound most of the Hudson Bay Basin, resemble ramparts of some gigantic fortress. This ridge is called the "Height of Land" on maps of Canada, such as Goode's School Atlas by Rand McNally Co., 1930.

This ridge of land is also a watershed. The rivers on the inside flow toward Hudson Bay, while those on the outside of the ring flow away from it. This "bowl" was created by the Hudson Bay Basin Ice Cap.

The lips of any bowl of earth which once held an ice cap must naturally appear as a height of land after the glaciers have melted and disappeared, because glaciated rocks and earth flow under pressure.

Mountain ranges occur along the eastern coast of the Hudson Bay Ice Cap Basin, bordering on Davis Strait—called Labrador Highlands and Penny Highlands in Labrador and Baffin Island, respectively. These all have appearances of being analogous to the mountain ranges along the coasts of Antarctica.

A peculiarity of the Laurentian Shield is the fact that the rock formations slope gradually on the inside of the bowl, toward Hudson Bay; but, the slope is steep on the outside of the bowl, in which direction the rock materials were pushed.

The outer edge of the Height of Land drops off 200 to 300 feet per lateral mile, and is fairly uniform for more than four thousand miles. The diameter of this great ring ridge averages about 1,500 miles, and the land area is shaped like semi-plastic mud would appear into which you had slowly pressed your booted foot.

Evidence of Careenings of the Globe

Such a ridge, ring, or "height" could be expected to develop in a natural manner if an irresistible force were pushing rocks and dirt; and it is therefore assumed, by analogy, that similar pressures and the same process of nature are now causing similar expansions and elevations in the lands surrounding the Antarctic Ice Bowl.

The rock floor of the Hudson Bay Basin Ice Cap—like that of Antarctica—was moved in from a tropical climate by a careen of the globe, and it was covered by tropical vegetation. Then, under ice pressures which rose to approximately four tons per square inch, caused by the ice masses which gathered above it, all forms of vegetation were compressed to an amorphous layer. Tree trunks and branches were deformed and obliterated by the ice masses which slithered over them. Fossil specimens of past growths of vegetation will rarely be found intact.

Telltale residue of buried organic matter, whose original forms have been entirely eliminated (as mentioned as organic materials in clay) are now being found in sections of the Laurentian Shield—especially in the black, slaty shales of some horizons of the Lake Superior region. They owe their color to the dissemination of carbon derived from organic matter.

An additional analogy is the Panama Canal, whose bottom was incessantly forced upward in bumps and bulges, due to isostasy and the weight of the surrounding hills. Lateral underground earth flows equalized the earth pressures when the lesser weight of water in the Canal replaced the heavier weight of the earth and rocks removed.

A communication from Hugh M. Arnold, Engineering and Construction Director, Panama Canal Company, Balboa Heights, Canal Zone, of May 10, 1955, states: "Any measurable heaving in the Panama Canal channel has been in the area of the famous Culebra slides, intricately slickensided and highly bentonitic. Pressure of the 'rock' from the adjacent banks sometimes results in a measureable bulging or heaving of the bottom of the excavated canal."

Risings or upswellings of the earth surrounding areas of

meadow land in the vicinity of New York City have often been noted following attempts made to reclaim such lands by filling with dirt.

Minor reasons for the growth of the Antarctic continent are the terminal moraines which are developed from the materials carried by the flowing glaciers and by the waters which in summer flow out from beneath the ice.

Some of the many small mountain tops—nunataks—observed near the coasts of Antarctica will eventually be found to be detached blocks riding shoreward on the glacial ice, as did the small mountains of northern New Jersey and southern New York which rode the glaciers during the Hudson Bay Basin ice age.

The geology of Antarctica shows that the land was successively submerged under the oceans, and was also repeatedly above sea level during previous epochs of time. The horizontal strata of many of the mountains contain layers of sandstone, limestone, granitic rocks, and coal. The older rocks are assumed to be at the bottom and the younger rocks at the top.

The limestone layers were created by corals and shellfish in shallow ocean waters. The sands of the sandstones were also created in shallow ocean waters. The granite rocks were created in upland areas (see "Origin of Granite" in Part III, page 228). The coal strata, found in the mountains, show that the coal areas were once wooded bottom lands, marshes or lakes, in tropical or temperate climates, where water-logged vegetation was prevented from oxidation and thus became coal.

The icebergs all drift away from Antarctica—just as floating apples, in a rotating tub of water, drift away from the center of spin. The rotation of the earth creates a throw of centrifugal force which causes the icebergs to move away from the Axis of Spin. Local southerly winds aid in starting the northward motion; but the icebergs continue northward after the local winds are left behind.

Nature is setting an example for us to follow. She is showing us how we, too, can delay the next careen of the globe by making the Antarctic Ice Cap lessen its weight by making it shed some of its ice. If we act within Nature's time limits, and if we

Evidence of Careenings of the Globe   117

greatly accelerate the throwoff of the icebergs, we can postpone the fatal day when the great Ice Cap will roll the globe sideways, and follow the icebergs to the tropics.

The great South Pole Ice Cap is a product of forces of Nature which are created by the "Will of God" and are beyond the understanding of men. The Ice Cap grows larger according to the Laws of Physics—which some of us must understand or most of us must perish. The Ice Cap must be subdued by man or man will be subdued by the Ice Cap. Like the Sword of Damocles poised by a hair at Dionysius' banquet, it threatens destruction, and in this case it will mean the destruction of most of the human race.

Our corporal salvation depends on our ability to control the further growth of the Ice Cap. Bleeding off the ice and making it gravitate through newly made channels to the coasts, can lead to our only salvation!

The North Pole area, in the epoch lasting until the next cataclysm of the earth, will not have an ice cap, unless Bering Strait should be closed.

Due to the Drag of Gravity (to be fully discussed in Part II), the waters of the Pacific Ocean are forced against the west coasts of North and South America. Because of this pressure, warmer sea waters flow constantly through Bering Strait, from the Pacific Ocean to the Arctic Ocean, at a normal speed of about four knots (about 4.6 miles per hour). Information regarding this rate of flow has been furnished by the U. S. Navy Hydrographic Office, which has since indicated that it may be only a surface speed. A daily flow of around 10 trillion cubic feet, or approximately 41 billion tons of water, was indicated by empirical measurements made by the U.S. Coast Guard in 1936.

The currents in the Arctic Ocean were discovered when the ship *Jeannette* was crushed by the ice and abandoned off Wrangell Island in 1881 and some of its wreckage was picked up three years later off the coast of Labrador.

The drift of the ice and the flow of the waters were proved to exist by Fridtjof Nansen during his voyage on the *Fram*, and also by a similar Russian vessel *G. Sedov*. Both drifted from a position near the Pacific side to the Atlantic side of the Arctic Ocean by becoming frozen in the floating ice. They both demonstrated that the surface ice flowed in the direction of the Atlantic side at a speed about the same as that of the wreckage of the *Jeannette*.

The floating ice on the surface of the Arctic Ocean is about 10 to 16 feet thick and it becomes rumpled into hummocks by the pressures created by winds and currents.

Snow falls on the drifting ice, which acts like a conveyor belt carrying it into the Atlantic Ocean, where both the snow and the ice melt.

Were it not for the continuous flow of water, from west to east through Bering Strait, the Arctic Ocean—which averages about 4,000 feet in depth—would have frozen solid centuries ago, and an Arctic Ice Cap would have developed.

With two Ice Caps—one in Antarctica and another in the Arctic—our epoch of time, between two careens of the globe, would have been radically reduced in length, and our present civilization would never have had a chance to develop.

Bering Strait—as a connecting link between the oceans—is seen to be of vital importance to our civilization: it makes it possible for the waters of the Pacific Ocean to flow into the Atlantic Ocean, via the Arctic Ocean, and thus acts to limit the accumulation of cold-storage ice at the North Pole.

II

MECHANICS OF THE GREAT DELUGE

THE earth is a freely suspended oblate spheroid. It spins on its axis, which is slanted 23° 27′ in relation to its orbit, and it makes a complete revolution every 24 hours. It moves around the sun and completes its orbit in a little more than 365 days. It has seven motions. Its four major motions are:

1) Rotation on its Axis of Figure, daily.
2) Revolution about the sun, yearly.
3) Careenings at irregular recurrent intervals.
4) Travel through space, with the whole solar system.

Its three minor motions are:—

5) Precession—a slow motion of the Axis of Spin in a cycle of time of about 26,000 years.
6) Nutations or noddings, which are slight, recurrent, variable, irregular motions during its precession.
7) Wobbling, which is a rotary off-center motion of the Pole of Figure around the Pole of Spin in a time cycle of about fourteen months.

The true axis of the earth is called the Axis of Spin; it can be considered as a line in space moving sideways in celestial space with the whole solar system. It coincides approximately with a line drawn from the North Star of the northern hemisphere to the Southern Cross in the skies of the southern hemisphere.

The Axis of Figure, on the other hand, is not permanent, but varies with each successive epoch of time. It is an imaginary line drawn through the center of the earth at its shortest diameter, and is determined by the location where the bulge of the earth happens to develop for any one epoch. This Axis is always *farthest from the center of the bulge.*

The two points where this axis meets the earth's surface are called the Poles of Figure. They are the North and South Poles of our geographical maps. This axis is not fixed in the same sense that the Axis of Spin is fixed in space. It is a geographical axis changing with each successive epoch. The areas of the earth surrounding the poles become centers for great snow and ice accumulation during each epoch of time—because of constant snowfall and lack of sun heat.

At the beginning of each epoch the Axis of Figure and the Axis of Spin coincide. At the end of each epoch they have become separated because the liquid materials which comprise most of the upper surface layers of the earth, have been redistributed. Great volumes of water are slowly withdrawn by evaporation from oceans and rivers, are precipitated as snow, and are solidified as ice at the poles.

In a telescope rigidly set and pointed at a fixed star we will see that the star sweeps by once in every 24 hours, due to the rotation of the earth. It has also been discovered that the star rises and falls in the telescope sight. The earth and the telescope have been found to be rocking slowly up and down.

The motion of a badly-thrown iron quoit gives an exaggerated idea of the earth's wobble. A point on the Equator may be thought of in terms of the iron quoit, and that point is found to be rocking north and south through a variable range. The maximum has not yet exceeded 50 feet, and the time period of the motion is about fourteen months. The true latitude of any point on the earth's surface varies to that extent. The well-known wobbling motion of a spinning top before it falls over on its side helps us to visualize the wobble of the earth.

Observatory records of the rocking motion of the earth are being kept continuously, at a number of different points on the surface of the earth but in approximately the same latitude. When the individual observations are grouped together it is found that the earth's poles are wandering in a circular motion.

Mechanics of the Great Deluge

The variations in the distance between the Pole of Figure and the Pole of Spin have been plotted on the assumption that the Pole of Spin remains in a fixed position. These observations show that the Pole of Figure is traveling around the Pole of Spin. The Pole of Figure moves from one side of the rotation pole to the other side, or, during some periods, completes its entire circular motion on one side of the Pole of Spin.

The observed motion of the earth's Pole of Figure in a period of no more than a few years has been charted. More recent graphs of the wandering of the Pole of Figure indicate that its circular motions have been completed entirely to one side of the Pole of Spin; this appears to indicate nutations in the wandering of the Pole of Spin as well as of the Pole of Figure.

The wanderings of the earth's poles prove that the whole earth wobbles: the wobbling * has the same effect on any point on the surface of the globe as it has on the Pole of Figure.

The discovery of the continuous wobble of the earth completely undermined and disproved a previously prevailing theory that any relatively rapid change in the position of the earth's Axis of Figure was impossible, the reason being the earth's tremendous gyroscopic energy of rotation. The wobble of the earth results in a wandering of the Pole of Figure, the center of gyration of the successive polar ice caps. The linear speed of the ice caps increases at about 6.28 ($2\pi r$) times their distance from the Axis of Spin; the total energy of motion and the throw of centrifugal force of the ice caps both increase proportionally with the weight of the ice mass that is off-center and at a rate which is the square of its velocity. The speed of travel is a function both of the rotation and the wobble of the earth. At the present time the speed of travel due to the wobble of the earth is extremely slow, less than 13 feet per hour, because its distance off-center is estimated not to exceed 50 feet. Nevertheless, as the ice cap grows in size the wobble of the earth imparts to it a greater and greater linear motion which is mirrored in

---

* For the cause of the wobble, see page 190, Part II—"The Drag of Gravity."

an increased centrifugal force, and this in turn tends to provide greater inertia.

These huge and transient weights at the poles develop a strong "throw" due to their eccentric centrifugal force of rotation, which overcomes the resistance to shear off the earth's materials—causing earthquakes and the forced migration of the weights to rotate under the sun of the heavens, instead of under the North Star and the Southern Cross.

We say that they have moved to near the Equator, but the Equator must be thought of as a theoretical line of latitude with respect to the sun, and not as a line fixed on the earth. Any circumference of the earth's surface would be at the Equator if it rotated farthest away from the center of spin.

We must also remember that the bulge of the earth about the Equator has no respect nor attachment for any particular continental land areas. Africa and South America happen, at the present time, to be parts of the bulge. This is a temporary condition, limited to our present epoch. During the preceding epoch Africa was at the North Pole and South America and North America lay in tandem along the Equator. A single careen of the globe accounts for that change.

The globe reels or careens, but it continues its rotation uninterruptedly on a new axis—in the same way that a billiard ball, which has rebounded from a table's cushion, continues to rotate but on a different axis.

When we compare the energy of the eccentric throw of the ice caps when rotating off-center with the energy stored in the rotating bulge of the earth, we must first realize that the bulge of the earth is not a rigid, solid ring, like the revolving element of a gyroscope, but that it is made up of yielding earth materials of rock and dirt that are relatively plastic when compared with the forces acting upon them. The bulge rearranges its elements and accommodates itself to these forces.

It is readily apparent that the energy stored in the rotating bulge of the earth is many thousand times greater than the energy of the eccentric centrifugal force or throw of the ice caps when rotating off-center. Many persons appear to have

Mechanics of the Great Deluge

misapplied this equation of forces in trying to prove that careenings of the globe are unthinkable and chimerical. Their error is one often made in nonanalytical studies. They have misapplied their mathematics by starting with the false premise that the bulge of the earth is rigid and solid, like a commercial gyroscope. The fact is that the earth changes the arrangements of its individual parts, including the elements that make up the bulge, when the earth's Axis of Figure moves away from the Axis of Spin.

The location of the bulge of the earth is fixed by the Axis of Spin and not by the Axis of Figure. If the Axis of Figure, at the South Pole, with its surrounding ice cap is moved ten miles north from the Axis of Spin, the equatorial bulge would then change its latitude and move ten miles south, because it responds exclusively to its rotation on the Axis of Spin, which remains relatively fixed. Therefore, this would cause a very great rearrangement of the rock and dirt elements that constitute the present bulge.

In the second section of this book it will be shown that the globe is held in its place in the celestial universe by static electrical repulsion, and that it is rotated by the forces of radiant energy which emanate from the countless billions of suns which form a part of the universe. Dynamic electrical radiations cause the rotation of the earth and the phenomenon of the weight of materials, and static electrical repulsion fixes the earth in its position and causes the ocean tides. It is the incoming, bombarding radiations which cause the earth to have a true axis or Axis of Spin around which it rotates in a position and a manner determined by these external forces.

It is thus cosmic forces of extraterrestrial nature that cause the earth to rush through space, revolve around the sun, rotate on an axis, and wobble. The development of the theory of "The Drag of Gravity," discussed in Part II, has become a natural by-product of the research required to establish the theory of the intermittent careenings of the globe and the recurrent cataclysms.

The bulge of the earth has been created and is maintained

by the external forces which rotate the earth and produce centrifugal force. Centrifugal force is the direct cause of the bulge, but this force is itself the product of the radiant energy of the external forces which impinge upon the surface of the earth. The bulge responds exclusively to these centrifugal and external forces and, therefore, it must always be centered very nearly along the line of the true Equator.

The centrifugal force which causes the bulge of the earth is a tremendous force, and yet it is relatively small compared to its opposing forces; otherwise the bulge would be much greater, since action and reaction are always equal and opposite in direction; and since the gravitational pressures from all sides are so very much greater than the tremendous centrifugal force which creates the bulge, the globe is nearly a perfect sphere.

The diameter of the earth at the Equator is 7,926.677 miles.
The diameter of the earth through the poles is 7,899.988 miles.
The difference between these two diameters is 26.689 miles.

(From *The World Almanac 1957*—quoting U.S. Coast and Geodetic Survey)

Based on the above we find that:

The mean diameter of the earth is 7,913.3325 miles.
The amount of the bulge of the Equator is approximately 13.34 miles.
The amount of the flattening at the poles is approximately 13.34 miles.
The maximum bulge on each side of the Equator is approximately 6.67 miles.
The maximum flattening at both North and South Poles is approximately 6.67 miles.

From these figures it is seen that the earth, though technically an oblate spheroid, departs from being a perfect sphere by about ⅙th of 1 per cent. The earth's diameter at the Equator, where the bulge is at a maximum, differs from the mean diameter by approximately .0017—too small a variation to show on a small school globe.

The bulge of the earth thus represents a definite arrangement of the earth's elements, adjusted automatically to definite centrifugal and gravitational forces, its general equilibrium being maintained by the yielding or flow of rock materials beneath the surface.

Were it not for its bulge the earth would be free to rotate about any axis or none. It would roll constantly; there would be no Axis of Figure. The extra weight of materials in the bulge establishes the earth transiently and precariously on a definite Axis of Figure.

The size of the bulge remains approximately constant throughout each successive epoch of time. But, during each epoch, ice caps continually grow at either or both poles. The bulge of the earth, then, can only temporarily ensure a dynamic equilibrium of the globe on its Axis of Spin; there is no alternative but for the ice caps to tip the earth over when they have grown too large for effective adjustments to be made through isostasy.

The opposing force which temporarily prevents the ice cap from starting toward the Equator is the stabilizing gyroscopic action of the earth's bulge. This opposing force, F, is transmitted through the constituent earth materials; the amount of force transmitted is limited by the resistance to shearing of the upper stratified layers of rock and hardpan, F.

The earth can continue to rotate on any one axis only as long as resistance F_1 is equal to or greater than force F. (Force F is a cotangent function of circularly rotating energy, expressed by $E = \frac{1}{2}MV^2$—where M is mass, V is velocity, and E is the energy developed.)

Referring once more to Fig. 1, the true axis, or Axis of Spin, and the true Equator are shown by solid lines. The North Pole, the South Pole, the Axis of Figure, and the actual Equator are shown by broken lines. The ice cap is assumed to be centered at the North Pole and its center of gyration is separated from the Axis of Spin by the distance R.

Two equidistant weight elements of the ice cap may be likened to the two balls of the well-known centrifugal "fly-ball" governor. These rotate in a 24-hour period of time around the Axis of Figure. The "fly-ball" at the left exerts a force which pulls the Axis of Figure toward the true axis. The one at the right is tending to pull the Axis of Figure away from the true axis and toward the true Equator.

The centrifugal pull of the two elemental "fly-balls" is the same on both sides of the Axis of Figure, but on the true axis the ball at the right exerts a much greater pulling force away from it, because it is gyrating at a longer radius by an amount equal to R. Similarly, the ball at the left exerts a smaller pull toward the true axis than its pull on the Axis of Figure because it is gyrating around the true axis on a radius shorter by R.

The algebraic sum of the centrifugal forces of all of the equidistant pairs of weight elements, rotating on the Axis of Figure as a pivot at a distance R from the true axis, is the centrifugal force which tends to throw the ice cap away from the true axis and toward the true Equator.

The centrifugal force of that part of the ice mass which is rotating in a 24-hour period of time, and out of coincidence with the Axis of Spin by a distance R, corresponds to the combined forces of all of the individual weight elements, and this is the eccentric "throw" which tends to move the ice cap away from the Axis of Spin. Each time that distance R becomes great enough an entire ice cap starts moving toward the true Equator.

F, the total force for any mass of ice rotating eccentrically at any assumed radius of gyration R, can be readily calculated, but little is known about the force of cohesion of earth materials, F_1, or how much resistance to shearing F_1 offers to the imposed force F.

But we do know that there is some distance for R which is critical, beyond which it is unsafe for the eccentric ice mass to go. Thus whenever the eccentric weight of an ice cap rotated at a radius one foot greater than the critical or safe radius R, then the force F overcame the cohesion of earth materials represented by F_1, and once that cohesion was overcome and broken down there was nothing to stop force F from increasing rapidly and wildly until the whole ice cap neared the true Equator and

Mechanics of the Great Deluge 127

became a part of the equatorial bulge of the earth—where it continued to pull eccentrically on the Axis of Spin but could not capsize the earth any further.

By considering an ice cap as a great horizontally mounted flywheel, it is possible to compute the centrifugal pull created by its "off-center" motion of rotation.

Let W = Weight of rim in pounds
R = Mean radius of rim in feet
r = Revolutions per minute
g = 32.16
v = Velocity of rim in feet per second—$2\pi Rr \div 60$

Centrifugal force of whole rim:

$$F = \frac{Wv^2}{gR} = 0.000,341 WRr^2$$

This force tends to disrupt one half of the wheel from the other half, and is resisted by the tensile strength of the flywheel rim.

In the case of that section of an ice cap which is rotating eccentrically about the Pole of Spin, its entire force tends to disrupt the earth—which may be likened to the flywheel rim—because the entire weight is gyrating like an unbalanced flywheel with all its weight concentrated at one point of the rim.

In the case of the eccentric elements of an ice cap, revolving once in 24 hours, or 1,440 minutes

r becomes 1/1440
F then equals 0.000,000,000, 164WR.

On the basis of this equation, the centrifugal force of a mass of ice of any weight W, at any radius of gyration R, can be calculated.

Suppose that the Pole of Figure and the Pole of Spin happen to be in coincidence, and that the ice at the pole is two miles high. Then, let us describe a circle with the pole as its center and with a radius of 50 feet. We will then have a cylinder of ice with its center at the Pole of Spin and weighing 424,560,000 pounds approximately, the weight of ice being assumed to be 57.5 pounds per cubic foot.

Now, let us suppose that the Pole of Figure and the center of the ice cylinder move, or wander, to a distance 50 feet from the Pole of Spin. Then the edge of the cylinder will be at the Pole of Spin and the whole ice cylinder will be rotating eccentrically "off-center" around the Pole of Spin and along a circle with a 50-feet radius. The distance R will then be 50 feet and, on this basis of the above equation, the centrifugal force F will be about 39 pounds.

For cylinders of ice of larger diameters—all assumed to be two miles high, and all assumed to wander from the Pole of Spin so that their edges are at the Pole of Spin and their centers at the Pole of Figure—we get the following values for force F:

R	F
50 feet	39 pounds
100 feet	312 pounds
200 feet	2,502 pounds
400 feet	20,022 pounds
One mile	46,050,000 pounds

These figures indicate the force applied as a sidewise heave or pull to make the globe reel and careen. The actual heave is greater, because for each elemental ice section of the mass that has moved away from the Axis of Spin, an equal elemental ice section, on the opposite side, has moved the same distance closer to the Axis of Spin. The centrifugal force of the one elemental section moving away is increased, while that of the one moving nearer, and pulling in the opposite direction, is reduced. The actual sidewise heave is therefore greater than the magnitude of the "throw" or heave arrived at by the flywheel formula.

A flywheel of any kind, anywhere, rotating out of balance, is a signal of danger to any engineer.

When a rotating gyroscope is suspended in trunnions, so that it is free to move in any direction, and a force F is applied to it at any point, the effect of the force becomes apparent at 90

Mechanics of the Great Deluge

degrees of a circle from the point of pressure. The application of a force causes a motion of the whole gyroscope at right angles to the direction of motion of its rotating elements.

To understand clearly what happens, consider yourself standing in space and that one element of the material at the Equator of the earth—the size of a rifle bullet—comes by you at a speed approximately 1,500 feet per second, and that you strike it on the side with a tack hammer as it passes by.

Due to the very great stored kinetic energy, or momentum, of the bullet you would not budge it, apparently, by the blow of the tack hammer; but you would deflect its course. You would change the direction of its motion. Instead of its motion continuing parallel to the Equator it would now travel at an angle to the Equator, but would continue around the earth in a great circle.

If we take a globe and fasten a string or rubber band around it at the Equator, that string will represent the path of the imaginary bullet. Now, if we deflect the angle of the string, as by the imaginary blow of the tack hammer, at the longitude of Greenwich, England, which is 0°, and we then straighten out the entire string, at the new angle, it will describe a great circle around the earth at an angle to the Equator.

The original motion of the bullet was from west to east. Therefore, if the tack hammer is delivered from the south, the string will lie north of the Equator from 0° to 180° longitude, and then will cross the Equator and lie south of it from 180° longitude back to 0°. Its maximum distance from the Equator will be to the north at 90° E. longitude and to the south at 90° W. longitude.

The maximum effect of the force is seen to be produced at 90° of a circle away from the point of application. The force, which was comparatively small, did not move the gyroscope, but it did change the angle of one of its rotating elements.

By this visual demonstration the motion of a gyroscope at right angles to the applied force is made clear. Now, instead of one bullet, consider billions of billions of bullets, and we have the whole bulge of the earth represented by bullets moving at approximately 1,000 miles per hour, or about 1,500 feet per second.

The pressures of the "off-center" centrifugal force of the present antarctic ice cap elements are apparently not being applied continuously at any one point of the gyroscopic energy of the rotating bulge of the earth, because the earth's axis is known to be moving. The Pole of Figure is wandering in a circular motion which is changing the point of application of the "off-center" centrifugal pressures; but a force applied continuously to a gyroscope will eventually bring about an appreciable change in the angle of direction of its motion.

The new order, signifying the next epoch, will begin after the careening and the cracking up have resulted in the equalization of the three forces:

1. The centrifugal force of the rotating ice cap
2. The centrifugal force of the earth's bulge
3. The forces of gravitational pressures

In order to visualize the force exerted on the earth by the eccentrically rotating ice cap and the manner in which that force tends to cause the earth to careen, think of a large flagpole to which a rope is attached at the top; it meets the ground at a distance of about 100 yards from the base of the pole. Some men pull on the rope and bend the flagpole. Now, let that flagpole represent the South Pole and you will get a mental picture of the force being exerted to move the Pole from where it is now located.

The analogy must be corrected by noting that the South Pole is not a flagpole, and that the force tending to shift the earth's axis is not applied externally, as with a rope, but is developed internally, by the centrifugal force of the eccentrically rotating ice mass. This force pushes the earth materials in front and pulls the materials to the rear, since all earth materials are held together by cohesion. The force is centered at the South Pole, which is the south end of the Axis of Figure, because the South Pole is the central point about which the ice cap has grown.

Mechanics of the Great Deluge

The "pull on the pole" exerts a pressure to cause the Pole of Figure to move away from the Pole of Spin and toward the Equator. The "pull" is always in a straight line from the Pole of Figure to the Pole of Spin, and the tension of the pull rotates around the Pole of Spin, with the ice mass, once in every 24-hour period. The force is not a constant, but varies with the weight of the off-center elements of the ice cap and with the speed of its motion, which increases with distance from the Axis of Spin.

When the Pole of Figure moves to a spot twice the distance from the Pole of Spin, the force, or throw, becomes four times as great as it was originally. The "pull" to move the ice cap toward the Equator increases at a rate equal to the square of the distance between the Pole of Figure and the Pole of Spin; simultaneously, the weight of an equal amount of ice on the opposite side of the Pole of Spin and its pull of centrifugal force in the opposite direction are reduced.

Another analogy might be a large spherical balloon with a cage or basket suspended closely below it, floating quietly in the air. Now, imagine that a sudden squall strikes balloon and basket and causes the basket to swing to a horizontal position compared to the balloon. The careening motion of the earth and its ice cap resembles the careening motion of the balloon and its cage.

As a third analogy, consider a globe two feet in diameter fitted closely but loosely inside a barrel hoop, and that the hoop represents the bulge of the earth at the Equator, even though the earth's bulge does not rise up so suddenly and sharply as the barrel hoop might indicate. Imagine that two nails, at opposite sides of the Equator, support the globe in the hoop—which is revolving with the earth of which it is a part.

Now, imagine that you are holding on to the hoop at the nail pivot points, and that the globe is out of balance and therefore starts to rotate on its pivot points, and turns a full 90°. This illustrates how areas, once at the Equator, move to the poles— 13 miles nearer the center of the earth—and how other earth sections move, like an underground wave, as they cross the relatively stationary bulge; and polar areas finally become a part of the new centrifugal bulge while moving 13 miles farther away from the center of the earth.

As a fourth analogy, consider the planet Saturn with its rings. Let us suppose that the rings are contracted within the body of the planet, and that they correspond to the centrifugal forces causing the earth's bulge. Let us then suppose that the planet itself careened ninety degrees or one-fourth of a full revolution, because of internal forces, but that the rings kept in their original position without careening.

We will then suppose that when the planet's materials passed through the rings that the materials were elevated as much as thirteen miles, and the rings thus wrecked the surface materials of the planet. That is the way the surface materials of the earth are wrecked when the earth careens, and the materials pass through the earth's bulge and then rearrange themselves again to conform to the centrifugal force of rotation which causes the bulge, and to the gravitational pressures which cause the earth to be spherical.

We find that animals have been tossed about (mammoths in Siberia) and torn apart (mammoths, bison, and other animals in Alaska), trees broken off at ground level and denuded of limbs (Yellowstone National Park, and elsewhere), and both animals and trees have been buried in the ground which later has become hardpan and rock.

The evidence indicates a super hurricane and dust storm where dirt, stones, trees, animals, and débris were lifted and floated by the air pressure and strewn about by the force of its motion. Effective resistance to the speeds of careenings is offered by the inertia of the earth's weight and the various kinds of surface resistance of earth, air, and water. Buried animals and vegetation give us clues as to the effects of such resistance to the motion of the air.

The resistance of earth materials and rock formations to the onslaught of the flood waters is indicated by the huge land areas that have been gouged out—leaving buttes, bluffs, and table mountains where the earth materials did not give way to

Mechanics of the Great Deluge 133

the floods; but the surrounding materials of rock and earth have been carried away and scattered elsewhere. We find ancient rock formations cropping out in areas which have borne the brunt of the floods, while in other areas—not assaulted by the rushing waters—similar formations lie deep in the earth.

The speed of careening varies with the locations. The maximum is at the Pole of Figure, the minimum at the transient Axis of Careen.

The speed of the ice cap accelerates rapidly during the first 45 degrees of latitude traveled, because then the centrifugal force of its motion would be pulling nearly sideways; but due to the spherical surface of the globe, this side pull changes to an upward and outward pull. By the time the ice cap has traveled 80° there is very little side pull, but there is a very great upward stabilizing pull, away from the center of the earth. The ice cap becomes a part of the new bulge of the earth.

It seems that the careening motion will stop when the new bulge and the ice cap work in unison to stabilize the globe on its new Axis of Figure. The kinetic energy stored in the careening continents will then appear to have been absorbed by the elevation of land areas, the creation of mountains, and the drift of continents—as previously pointed out.

The existence of the Land and Water Hemispheres and of the Great African Rift—just as they are presently conditioned—fits accurately into the careening globe theory. It also confirms the theory that the Sudan Basin was at the North Pole of Spin during the previous epoch of time, and it confirms the existence of an ice cap which traveled to the present location of the Sudan Basin—where the ice melted and left its mark by irregularly radiating, gouged-out water courses which are now dry land.

Following the next careen of the globe the present continent of Antarctica can reasonably be expected to become the center of a land hemisphere—because of the centrifugal force of rotation which will be created by its weight and speed of motion. This transient force will not only pull Antarctica but also its surrounding ocean floors upward and keep them above sea level, thus creating new land areas.

The area of the globe now occupied by the Arctic Ocean will probably become the center of a Water Hemisphere—like the Pacific Ocean today. What is now northern Siberia, northern Canada, and Alaska, will probably become parts of the submerged ocean floor.

We may learn when the next great deluge of the earth will normally occur if we can determine how much time has elapsed since the latest World Flood, also how long the past epochs of time lasted. This information may be obtained by counting the varves of growing clay beds, and checking the time element there disclosed against similar counts of varved clays of earlier epochs; we can also analyse the speeds of retreat of cataracts and the lengths of the river gorges which they have cut during their slow, creeping retreats; we can investigate the thicknesses of the average strata of earth of our epoch of time and compare them with average thicknesses of the average strata of previous epochs; and, we can check the area covered by our present South Pole Ice Cap against the areas covered by previous ice caps.

The year 1966 has been converted below to the corresponding year in six different calendars. These calendars were established by men whose work represents the most enlightened estimates of their time, especially as to the beginning of their eras:

	Year
Grecian Mundane Era	— 7,564
Civil Era of Constantinople	— 7,474
Alexandrian Era	— 7,468
Julian Period	— 6,679
Mundane Era	— 5,974
Jewish Mundane Era	— 5,727

The epoch of time in which we now live has run approximately 7,000 years since the last Great Cataclysmic Deluge of the earth. (The calendars of men begin farther back than 7,000 years.) Our epoch will doubtless last longer than the average epoch, because we have been fortunate in having only one polar ice cap and not an ice cap at each pole. Our most accurate time

Mechanics of the Great Deluge

scales are the river gorges—such as those of the Niagara and Mississippi Rivers—which have been previously discussed, and the counts of varved clays.

A four-million square-mile dent was left by the Sudan Basin Ice Cap. Glacial markings in North America cover about four million square miles; but the area within the lips of the Hudson Bay Basin is about two million square miles. Antarctica covers approximately 5,500,000 square miles; but we have no accurate data, as yet, regarding the dimensions of the bowl containing the ice cap.

In view of the evidence to the effect that the last flood occurred in the fall, it seems safe to say that the next World Flood will occur in the spring in the Northern Hemisphere. The ice cap that caused the last Flood was at the North Pole. The present epoch-ending Ice Cap is at the South Pole; thus, the seasons for the floods will be reversed.

Summary of evidence indicating that the last Flood occurred in the fall:

1. Mammoths that lived in or near equatorial areas of the earth, and now preserved in Siberia by quick-freeze, are found with short, thick undergrowths of hair. Many animals develop winter growths of hair or fur; in nontropical animals such growths generally are indications that autumn has set in.

2. Grasses found in the stomach of the Bereskovka mammoth had seeds; this is a usual indication of autumn.

3. The mammoth tree—found in the Siberian tundra had fruit on it, indicating summer or early fall, if it grew in a temperate climate.

III

Man: The Past, The Present, and The Future

The Survival of Civilization

MEN of science have discovered and collected records showing that human beings have existed on the earth for a half million to a million years. They have identified Java Man, Neanderthal Man, and other anthropological specimens, which they date back a very great number of years.

Our present civilization has developed during our present epoch of time, approximately 7,000 years. This epoch compares to the age of the earth—4½ billion years approximately—as one tick of a clock compares to all the ticks of the same clock running constantly for approximately seven days, with one tick for each second.

Keeping in mind these tremendously different time scales, we find it especially significant when qualified students of ancient history inform us that the historical records of the earliest civilizations—such as Assyria, Babylon, and Egypt—arose about 5000 B.C., and that prior records are legendary.

Authentic history concerning groups of human beings, therefore, began about 7,000 years ago. We know very little about the history of man prior to that last great Flood which destroyed most of the inhabitants of the earth. Our knowledge of history is limited to probably less than one per cent of the time that men have inhabited the earth.

Through migration the descendants of those who escaped destruction by the Flood have spread over the surface of the earth; most of the populations of Europe and America came from

The Past, the Present, the Future

the east; and the migrations of peoples and the histories of nations attest to the fact that these land areas and continents are rather new, and that it is only in fairly recent times that they have become established in the locations that we know today.

The equivalent of our present world population of over two billions of people could be created in a little over 600 years if one male and one female survived after a world deluge and had six children, and each pair of children had six children during nineteen generations. The nineteenth generation alone would number over two billion.

There are three places, possibly four, on the earth's surface which were not submerged in water during the last Great Deluge. They are areas which, during Epoch No. 1 B.P., were at the North Pole, the South Pole, and at the two pivot points on which the earth turned as it careened—the points which correspond to the poles of that transient axis.

The North Pole land area, which then comprised most of what is now Africa, careened until the center of the ice cap—which had been at a latitude of 90 degrees—arrived at a latitude of about 10 to 15 degrees, or approximately at what is now Lake Chad in the Sudan Basin of Africa.

The centrifugal force of that reeling ice cap pulled the land out and away from the center of the earth and, together with the centrifugal force of the spinning earth, left those land areas about 13 miles farther away from the center of the earth than when it was at the North Pole. As previously explained, the force of that pull was so great that most of the land areas of the earth are in a "Land Hemisphere."

During the careening, the centrifugal force of the Sudan Basin Ice Cap apparently caused the land to rise above sea level, for the adjacent land of Egypt appears to have escaped the Flood. Records seem to indicate that animal life there was not extinguished.

The South Pole Ice Cap area of Epoch No. 1 B.P. also reeled to the latitude where the old bulge of the earth had existed, and to approximately 13 miles farther from the center of the earth

Fig. 5. Approximate positions of land areas during the previous epoch of time. Alaska and the United States of America were in a tropical climate. The Sudan Basin of Africa was at the North Pole. X marks the temporary geographical location of the North Pole for our present epoch.

than when it was at the South Pole. This land area may have become submerged.

The survival of animal life there is not so clearly established as in the case of the North Pole Ice Cap region—principally because islands alone mark its approximate location.

It seems probable that the short duration of Epoch No. 1 B.P. was due to the existence of two ice caps, and that the North Pole Ice Cap was the larger, because its centrifugal force of rotation brought about the greater elevations now found in the Land Hemisphere.

The two pivot areas of the earth were not submerged. They were located on the earth's bulge at the Equator before the earth careened, remained on the bulge during the great Flood, and are now on the bulge of the new Equator. Except for tidal waves, coming from disturbances in other areas, these sections of the earth's surface escaped the Flood, a fact which explains why animal life appears not to have been extinguished in the two pivot areas.

The evidence for the continuance of animal life in the former North and South Pole areas, and in the pivot point areas on which the globe careened, will be discussed below.

The Past, the Present, the Future

THE FORMER NORTH POLE AREA included Egypt, which borders on the Sudan Basin, and in that country originate some of the earliest records of our present civilization. Egypt escaped the Flood, for at the "dawn" of Mediterranean history the nation appears to be mature, old, and entirely without mythological and heroic ages—as if the country had never known youth. At the time of Menes, the first king, the Egyptians had long been architects, sculptors, and painters.

Relatively speaking, the civilization of ancient Egypt, upon its first appearance, was of a higher order than at any subsequent period of its history—a fact which indicates that it drew its greatness from a fountain higher than itself or at least its equal. The civilization of old Egypt did not go through a period of infancy; it was mature when it first appeared on the stage of history.

The legend of the great Flood was not current in historic times among the Egyptians nor among the black races, according to Ignatius Donnelly, but it occurs almost everywhere else. The Assyrians and Babylonians had traditions of a Flood. Abraham migrated from Ur of the Chaldeans, near the head of the Persian Gulf. His grandson, Jacob, with his twelve sons, moved farther westward into Egypt, so that the Egyptians of that period must have heard about the Flood, though they seem to have escaped it.

The Egyptians were doubtless indigenous; the ancestors of Abraham arrived in Mesopotamia by boat. The records of the Egyptian priests antedate the Flood. The priests of Memphis reckoned back 11,000 years. The priests of Thebes told Herodotus that they reckoned back 17,000 years.

The Egyptians appear to have been living on the edge of a glacial continent during Epoch No. 1 B.P. The center of Egypt's earliest civilization was about as far from the North Pole of Epoch No. 1 B.P. as Fairbanks, Alaska, is from the present North Pole. This cold climate could account for the absence of many vegetables and fruits in early Egypt—such as the fruits and vegetables, listed presently, which have been found to have survived only in the pivot point areas of the earth.

An interesting but mythological clue to the land of Egypt

being related to a region to the north and to the land of the early contemporaneous Sumerians, a pre-Babylonian civilization in the Euphrates Valley, which suffered a flood, is referred to by an eminent scholar: "The Osiris story was the best known and most influential of the Egyptian myths. He introduced agricultural, animal husbandry, and arts and crafts to Egypt After his death he returned to his original home, from which he had presumably brought agriculture, animal husbandry, and arts and crafts to Egypt. Osiris's home was not in the west, as is usual with such happy hunting grounds, but in the north. The land was foggy and bordered with high mountains . . . Toward the mountains rose a dense forest . . . Many of the trees were conifers, sacred to Osiris.

"A similar mystery surrounds the origin of the Sumerians, the only people known to have had writing as early as the Egyptians. They brought their language into Lower Mesopotamia . . . after a flood of the twin rivers had wiped out some of the more vulnerable settlements of the earlier inhabitants. . . . The date of the flood is estimated at about 3000 B.C." (*The Story of Man*, by Carleton S. Coon pages 238–9.)

From the Near East four historical traditions have come down to us concerning the earliest inhabitants of that region. The Babylonians held to a tradition that they were descended from persons who survived the latest great deluge which destroyed everyone else. Their ancestors had survived in an Ark, which had come to rest on the mountain of Nizar—thought to be in the vicinity of Teheran, Iran.

The Greeks claimed descent from Hellen, son of Deucalion and his wife Pyrrha, who survived a great Flood which destroyed all others. They landed on Mount Parnassus in an Ark or chest. The Jews have given us the story of Noah, who with his wife, three sons and their wives were carried safely through a great Flood which destroyed all others. They landed on Mount Ararat in present-day Armenia.

The Koran informs us that Noah and other believers were the only ones to survive a great Flood. The Ark in which they were saved came to rest on Al-Djoudi. Noah's son perished in the great waves, because he was too late in embarking.

Remains of a vessel—declared to be the Ark—were found on Mt. Judi, on the left bank of the Tigris River, in the eighth century A.D. The remains were exhibited in a mosque and monastery of commemoration and were viewed by travelers and visitors from far away. The buildings were struck by lightning and burned to the ground in 776 A.D.

Each of these historical traditions is believed by many millions of differently educated people. Assuming that all four are true, then modern man must have had a great number of Flood-surviving ancestors in the Near East.

THE FORMER SOUTH POLE AREA   This region approximately encompasses the present Samoan Islands in the Pacific Ocean. Many Polynesian peoples have traditions to the effect that the center of dispersion of the race over the Pacific Ocean from Hawaii to New Zealand was Savaii, which is the largest of the Samoan Islands. Its highest elevation above sea level is now about one mile. Historians have remarked on the absence of any Flood tradition among the Polynesians of Oceania.

The Transient Axis of Careen

THE poles of this axis were the pivot points on which the globe careened. They were on the Equator before the latest careen of the globe and remained on the Equator after the careen. The compass points changed. The climate remained unchanged.

THE WESTERN PIVOT POINT AREA   is now known as Peru and Ecuador, in South America. It was a Safe Area for peoples and civilizations. Their descendants included the Incas, who continued to live in a mountainous region, usually not a natural place for a civilization to start.

The Inca civilization—like Egyptian civilization—appears to

have drawn its early greatness from a source perhaps higher than itself, for numerous basic food plants, on which previous civilizations no doubt depended, have been first discovered as growing in or nearby this pivot area. These same vegetable species were obliterated by the Flood everywhere except in pivot point areas.

Maize (corn), white potatoes, tomatoes, and certain beans—including Lima beans—were all found originally in this area and nowhere else in the world. They are indigenous to Peru, Ecuador, Bolivia, and northern Chile. Sweet potatoes, squash, and peppers were first found in nearby areas, including Central America, Mexico, and the West Indies.

Seeking traces of some of these vegetables in the opposite Safe Area, or eastern pivot point, where pre-Flood civilization was similarly not destroyed, we find that cobs and kernels of popcorn—identical with the corn found in graves in Peru—have recently been found among the backward peoples in the Naga Hills, where Burma meets Assam.

It could be glossed over as no more than a coincidence that maize has been discovered as an original food in both pivot point areas of the latest careen of the globe, were it not for our previous assumption that not only maize but also others of the western pivot point area vegetables would be discovered in the eastern pivot point area.

THE EASTERN PIVOT POINT AREA, now known as Sumatra and the Malay Peninsula, was a Safe Area. This fact accounts for the civilization of Asia being older than that of Europe. Europeans have generally considered some place to the east as having been the birthplace of the human race. Migrations of people have continually come from the east, pushing westward.

Most of our fruits came to Europe from the direction of the Malay Archipelago. Apples, pears, cherries, figs, olives, and most of our plums and grapes have been traced to the eastern Mediterranean and to the Caspian Sea areas. Peaches, apricots, bananas, mangoes, oranges, and lemons have come originally from regions farther to the south and east. Much botanical evidence points to Malaysia as the ancestral home of the coconut—though its definite origin is still in some doubt. (Authoritative accounts of the origins of our vegetables and fruits are contained in *The National Geographic Magazine.* For greater details, see "Our Vegetable Travelers", in the August 1949 issue, and "How Fruit Came to America", in the September 1951 issue.)

Five distinct races of men are generally recognized by anthropologists—a race being the descendants of a common ancestor, or family, tribe, people, or nation believed to belong to the same stock or breed. These five races are the Caucasian or white race, the black, yellow, red, and Eskimo. There are also distinct subdivisions—such as Pygmies, Bushmen, etc.

The careening globe theory may aid in identifying more clearly why these races are found where they are now. The Egyptians, for example, lived in a cold climate and were moved to a tropical climate, where they flourished.

The Eskimos of today probably lived in a tropical climate during Epoch No. 1 B.P. They are now in a frigid climate and have survived because they have managed to adjust to their environment. The mammoths of that same period perished when moved to the arctic regions, being unable to adjust to the new climate.

The Eskimos might become as prominent in the next epoch as the Egyptians were at the beginning of our present epoch.

The Migration of the Present Ice Cap

WHEN the earth careens due to the combined centrifugal pressures of the Antarctic Ice Cap and the Greenland Ice Cap acting along the same Great Circle, the event will be heralded by earthquakes along the Equator at 45° E. longitude, which is just off the African east coast and at 135° W. longitude, which is in the Pacific Ocean.

Those are the locations of points which are 90° distant, east and west, from the assumed points of application of the eccentric ice cap forces which are tending to change the angle of direction of the motion of the gyroscopic bulge of the earth. Those are the places where the maximum effect of these pressures would occur. The result of the pressures, if continuous, would be to produce a movement in the earth's strata at one or both of those locations.*

The earth gyroscope differs from a commercial gyroscope. Its internal parts adjust themselves automatically to the pressures exerted on those parts. Therefore, as earth parts yield to pressures, break down, and readjust themselves, a different careening motion of the globe occurs than is apparent in the commercial gyroscope.

The careening will not occur at a 90° distance from the point of application of the force, because the energy of the rotating earth bulge, against which the energy of the centrifugal force of the off-center ice cap is applied, changes its direction of motion and its geographical location as soon as the bulge starts to move.

The initial start of the careening of the globe will result from a yielding of earth materials along the bulge, near the Equator, at a 90° distance from the point of application of the force.

Once this change of location of a part of the bulge has started, the predominant force will be the ever increasing centrifugal force of the off-center elements of the ice cap, which increases at a rate equal to the square of the speed of motion, and which, in turn, increases directly with the increase in distance off-center, plus an increment due to the lessening pull in the opposite direction by the ice mass on the opposite side of the Pole of Spin as the ice cap moves sideways across the Pole. Thus, an enormous and rapidly increasing force is developed which carries the ice cap rapidly away from the Pole and toward the Equator.

Due to the curvature of the earth, the "throw" toward the

---

* This is discussed later on under "The Four Safest Areas for Animal Life."

The Past, the Present, the Future

Equator will lessen as the distance from the Equator lessens. Meantime the force of the kinetic energy of the entire ice cap will have taken over as the predominating force, and it will become blended with the kinetic energy of the rearranged materials of the new bulge of the earth into one gyroscopic force.

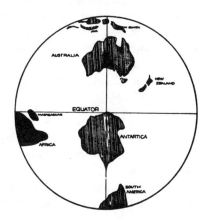

FIG. 6. The earth after the next careen, Antarctica area. Careen assumed on 135° 6 E. longitude. The antarctic Ice Cap has moved to near the Equator and this hemisphere will become the land hemisphere, with emergence of more land.

The center of the ice cap will reach a point on the new bulge within about 10 to 15 degrees of latitude from the Equator. Distance R will have become roughly 5,500 miles and, due to the increased speed of its motion, the kinetic energy of the rotating and reeling ice mass will have been increased to billions of times what it was at the start.

The start of the ice cap on its migration from the South Pole Axis of Spin to a point near the solar Equator, will occur for reason No. 1 (below), which may be influenced by reasons No. 2 and No. 3.

1. The ice cap's migration will start because of a yielding of earth materials, somewhere on the earth's bulge, due to a change in the angle of rotation of the earth gyroscope, responding to the pressure of the centrifugal force of the ice cap. The first yielding will occur at a place which will become a pivot area when the earth careens, as the previous explanation of the gyroscope indicates.

At the start, the ice cap is just enough off-center to exert a force sufficient to cause earth materials to yield somewhere. An earthquake occurs. If the earthquake does not relieve the pressure, it becomes cumulative for the reason that a crack-up of the earth in one place removes that area from the front line of defense, and throws the burden of resistance on other areas, while any yielding of the earth which permits an increase in the distance between the true axis and the center of gyration of the ice cap, increases, in proportion to the square of the distance, the throw of centrifugal force of that part of the ice cap which is not counterbalanced by the throw of the ice mass on the opposite side of the pole.

2. It may start because changes in the isostatic balance of the earth—due to shifts of materials elsewhere on the globe—are of sufficient magnitude to throw the Pole of Figure destructively off-center.

Normal changes of the isostatic balance of the earth occur slowly, due to such rearrangements of materials as:

a. Accumulations of glacial ice
b. Changes of shore lines resulting from lowering of ocean levels
c. Diversions of water courses
d. Impounding of water in artificial reservoirs
e. Shifting of sand and dirt by the winds
f. Growths of forests and bogs
g. Concentration of weights of materials in city areas
h. Extraction of oils and minerals from underground

3. It may start because of changes in the directions and intensities of the forces of celestial radiation, whose impact on the earth's materials cause its rotation. Minor changes produce a slight wobbling motion of the earth, as explained in Part II. The centrifugal force of the earth's bulge readjusts the Axis of Figure to these changes.

Some disturbances of the heavenly bodies most closely associated with the rotation of the earth—such as the near approach of a very large comet—might cause the position of the earth's

The Past, the Present, the Future 147

Axis of Spin to change too rapidly. A sudden change in the directions of the driving forces which rotate the earth, could move the Axis of Spin so far out of coincidence with the Axis of Figure that the stabilizing influence of the centrifugal energy of the earth's bulge would be overcome before it could readjust the Axis of Figure to the new position of the Axis of Spin. The larger and heavier the polar ice caps become, the more difficult these readjustments become.

The migration of the ice cap from pole to equatorial bulge will be marked by a slow start, by a rapidly increasing acceleration for the first half of the journey, and by a slowing down to a full stop before it reaches the Equator.

Whatever the cause of the migration of the ice cap, the consequence which follows is that the earth careens. The direction of its travel will be along a longitudinal meridian line—with a slight veering to the east, caused by gravitational pressures.

The last stage of the migration of the ice cap from the Pole to the Equator of Force* will be marked by a slowing down and final cessation of all latitudinal motion. It will become a part of the new equatorial bulge. Its speed will be that of the bulge, and the centrifugal forces of the ice cap and of the earth's new bulge will be merged into one centrifugal force.

The elevation of the ice cap will be approximately 13 miles farther from the center of the earth than when it was at the Pole. Two areas of the earth which were near the old Equator will become the new North Pole and South Pole. They will each be approximately 13 miles nearer to the center of the earth than when they were near the Equator.

To an observer at some point on the deluged parts of the earth it would appear that the bulge was moving. The bulge would appear to move like a huge underground wave or earthquake, as it remained in its true position of force at the true

---

*EQUATOR OF FORCE: term devised by author, meaning, as the earth careens, the bulge of the earth changes latitude about 80 degrees. At any particular time during the careen, the Equator of Force is located at the *bulge of force.*

Equator and the lands and seas passed across it. If, however, the observer looked up at the sun, moon, and stars instead of watching the earth, he would see that he was moving with the earth, and would feel that there was a rapid change of climate.

If the observer is standing at one of the pivot points of the careening earth, he will find that the compass points will change. The sun will strangely appear to rise in the south and set in the north, at one pivot point—because the land he had known as being south of him would be to the east. Similarly, at the other pivot point the sun would appear to rise in the north and set in the south.

Herodotus was told by the priests of Memphis there had been "341 kings and 341 high priests in 341 generations during 11,000 years, and in that space, as if to corroborate their genealogy, the priests asserted that the sun had risen twice where he set, without affecting any changes in their climate or the productions of the country." This reference to two changes in the relative geological points for the risings and settings of the sun in ancient Egypt, fits perfectly into the pattern of the careening globe theory, though failure to note some change in climate casts a doubt.

In the tomb of Senmut, the architect of Queen Hatshepsut of the XVIII Egyptian Dynasty, there is an astronomical ceiling panel which shows the Orion-Sirius group of stars proceeding in a direction opposite to their present motion. This panel could well have been made to commemorate an earlier historic period when the stars of Egypt did appear to be going in the reverse direction, and, if so, would confirm that east and west were actually reversed, as related by Herodotus.

Assuming, for illustration, that on the day of the latest careen of the globe, it was 9 o'clock in the morning in Memphis when the globe began to careen sideways while continuing its normal west to east rotation; people would then have observed that the sun appeared to stand still or to wander erratically, and to finally set near where it had risen—then considered to be the east. During that day, possibly before 6 o'clock, the east and west had changed places. Since people did not know that the

The Past, the Present, the Future 149

earth was rotating and careening, they considered that they had seen the sun stand still and then move backward.

This is all simply explained. The Sudan Basin was at the North Pole. Lake Chad was the central point. The ice cap rolled around to within 10 to 14 degrees of the Equator. The land of Egypt had been south of Lake Chad, and the sun had been rising in the east. But when the Arctic Ocean area moved to the North Pole, and Lake Chad moved to where it is now located, the sun of Egypt appeared to be moving in a direction nearly opposite to the usual one.

There are legendary records of unduly long and short days and nights in other parts of the world, as well as accounts of the sun and moon standing still.

Egypt's escape from inundation during the latest great deluge has already been discussed in the section on "The Survival of Civilizations." The energy needed to raise and lower continents and ocean bottoms was furnished by the kinetic energy of rotation of the entire earth. Some habitable land areas became submerged and some ocean bottoms became inhabitable land areas.

During the period of the next great deluge of the earth, a general chaos involving readjustments of land and water areas will take place. Some existing mountains will be raised and some lowered, compared to the level of the oceans. New mountains will be made.

The ice cap will start to melt at once under a blazing tropical sun. A dent in the earth will mark its former site. Dry water courses will mark the area as a telltale of the run-off of enormous volumes of water from the melting glaciers. Striations in the rock floor will show the grooves cut by the moving ice.

During the time required for the melting of the glacial ice, numerous earthquakes will occur as the ice mass along the Equator is reduced in volume, and the materials of the earth readjust themselves, in their new positions relative to the earth's centrifugal force and gravitational pressures.

The earth will again become stable. Animal life and activity will carry on at some or all of the four areas to be described later as being least affected by the careening of the earth and the corresponding deluge. Earthquakes will become less frequent as the earth's materials harden in new formations and dirt becomes hardpan and rock.

The depths of the oceans will increase until the ice cap has melted and until it again has become water. While the old glaciers are melting the two new polar areas will begin to grow their own ice caps. After the old glaciers have disappeared the ocean levels will again gradually be lowered by the evaporation of water which condenses and falls as snow and thus becomes a part of the ice caps at the new poles—and thus more land will gradually become inhabitable.

The cycle of the careening earth with its corresponding Great Deluge will be repeated again and again at the end of one epoch after another.

The Four Safest Areas for Animal Life

ASSUMING a hypothetical case of the globe careening due to the combined eccentric centrifugal forces of the Antarctic Ice Cap and the Greenland Ice Cap pulling in unison in the same direction—*i.e.*, along the same meridian circle—and that Greenland will start southward along approximately 45° 0′ West longitude at the same time that Antarctica moves northward along the continuation of that meridian circle, or 135° 0′ East longitude, then it may be conjectured that the four safest areas during the deluge which will accompany the impending careen of the earth will be Greenland, Antarctica, and the two pivot areas of the equatorial axis on which the globe careens.

The eastern pivot point of the transient Axis of Careen will be approximately where the meridian circle, East longitude 45°, crosses the Equator, in the Indian Ocean. That point is so close to the east coast of Africa as to warrant the expectation of the survival of animal life there, especially in the highland areas near the coast.

Greenland and the adjacent islands are expected to remain

The Past, the Present, the Future 151

inhabited by animal life, if the centrifugal force of rotation of the careening globe keeps those land areas above sea level, while their positions on the globe change from being near the Arctic Circle to being near the Equator.

Supporting this theory is the fact that the Antarctic ice mass would tend to develop an off-center ice weight, whose eccentric centrifugal throw would be approximately in the same direction as the centrifugal throw of the Greenland ice mass, the reason being that the geographical center of the antarctic ice mass and continent is about 5 degrees, or 350 miles, from the Pole of Spin and is located at about 80 degrees East longitude. According to this hypothesis, Brazil would roll around to the South Pole and the Philippine Islands would become the land area nearest the North Pole.

It would be an equally valid speculation to say that some area of the globe within about 2,000 miles of Lake Chad will be at the North Pole during the epoch of time following our own, and that this would occur as a result of the past—namely the Hudson Bay Basin careened to the North Pole Axis of Spin, then Lake Chad moved in, only to be supplanted by the present Arctic Ocean area. This shows a tendency for land areas to roll back to nearly the same position of latitude and longitude that they rolled away from.

Coal and oil fields, fossil forests, and the Canadian ice ages contain records of the fact that they were built up in tiered layers of the earth's strata—one below the other—this confirming a repetition in the directions of careenings. Also, we have found from a study of core drillings that the globe has a tendency to repeat the same kinds of materials in alternate strata of the earth.

Repetitive careens of the globe, back and forth, resulted in the telltale successive locations of former North Poles' "ice bowls" in the Caspian Sea Depression, the Hudson Bay Basin, and the Sudan Basin of Africa. However, something different may be expected to result when the South Pole Ice Cap takes over the role of initiating the roll-arounds of the globe. An entirely new sequence of global roll-arounds may be initiated. This section, therefore, has contained speculations and not precise forecasts.

An Exhortation

WHEN we were children we learned that in our daily lives we must conform to the Laws of Nature—or die. Now the time has come for us to similarly apply our knowledge of the Laws of Nature to the mechanics of our whirling globe, so that life on our earth may continue uninterrupted.

We must make use of some of these forces of Nature to prevent and overcome the baneful effects of other forces of Nature which—if not brought under control—will prove to be destructive to all life. The closer we study Nature, the closer we get to Nature's God, and begin to understand the causes and effects of observable phenomena.

This book has attempted to establish the truth about a coming great deluge of the earth, another in the series of recurrent careens of the globe, which have been occurring for a much longer period than man has inhabited the earth. This book also tries to show how the impending deluge can be postponed by the timely use of the power of applied knowledge. It clearly indicates the peril and just as clearly shows how and why a recurrence of these universal calamities may be postponed, and possibly avoided, by a concrete demonstration of man's will to escape destruction.

Awareness of the gigantic power to destroy inherent in the enormous unwieldy weight of the gyrating Antarctic Ice Cap, must be the first step in creating a cooperative group reaction to the deadly peril. People of education and initiative must become awakened to full awareness of the lurking danger represented by this wanton titanic power which is ready, able, and destined to end our civilization—if left uncontrolled by man.

It is, therefore, important that the facts be communicated

so that the menace will be generally understood and discussed. An awakening to the danger among the people at large is the first requisite!

The impending deluge should be a common topic of conversation, like the weather. "Everybody talks about the weather, but nobody does anything about it" was a standard joke a few decades ago, but since then a great new industry of air conditioning or "weather making" has developed. If everybody talked about the careening of the globe, as they do about the weather, then a great many people will try to do something about it. It will indeed become a matter of great personal interest to many people.

Let therefore all nations unite in pooling all their resources of atomic energy, mechanical equipment, and man power. Let us not continue to waste our substance in building giant mechanisms for destroying each other, while the growing Ice Cap is developing its latent resources to annihilate all of us.

Let there be an international war in which all nations fight as brothers against the common enemy. Let us attack the Ice Cap.

A Method

OUR OBJECTIVE must be to maintain the stability of the globe on its present Axis of Figure, by controlling the further growth of the Antarctic Ice Cap. Such arrangements must be made that eventually the annual flow-off will be equal to the annual accumulation of the glacial ice.

It is self-evident that the releasing of icebergs is the present safety valve which postpones the onset of the next great deluge. The waters of the earth tend to accumulate as ice at the poles, and if no icebergs were created the present world equilibrium could not long continue. Thousands of icebergs are cast off each year, their sizes at times exceeding two hundred miles in length.

The attack on the Antarctic Ice Cap is an engineering problem; the attack must be made along the 16,000-mile-long coastline and by men prepared to encounter and overcome very great difficulties. Controlling the size of the Ice Cap must begin with a great effort whereby more and bigger icebergs will be released annually.

The South Pole is on a plateau approximately two miles above sea level. The glacial ice flows northward toward the sea in all directions from the highest elevation. The distances of travel vary from about 600 miles to 1,900 miles, and average about 1,400 miles.

There is very little information available concerning the speed of motion of the antarctic glaciers. Some Greenland glaciers move at average speeds of over 100 feet per day during certain periods of the year. The gradient—or average slope—of Greenland from center to sea is roughly 23 feet per mile. The corresponding gradient for Antarctica is about 7 feet per mile.

If some of the barriers of rock, or mountain, which retard the flow of the ice to the sea, can be removed by engineering means, the natural flow-off of ice can be accelerated. Engineering research is thus seen to be one of our first requirements. We must find answers to the following and many other questions:

What is the annual precipitation of water (snowfall) in Antarctica?

How fast are the glaciers flowing to the ocean?

Can their speeds be accelerated?

What is the total weight of the flow-off of icebergs, annually?

How deep is the ice.bowl?

How wide is the ice bowl?

Which barriers are removable (so as to concentrate on the easier ones)?

Which, if removed, will produce the maximum acceleration of the backed-up ice masses?

What is the estimated cost of removing each of these barriers?

When we learn the present ratio of snowfall to flow-off, evaporation and ablation, we will know how much the flow-off must be accelerated to secure stabilization of the globe on its present Axis of Figure.

Since the discovery of the atomic bomb the detaching of, or

The Past, the Present, the Future 155

gouging channels in, the peripheral rock masses of the antarctic coastline—to permit the natural flow-off of the glacial ice—does not appear to be as formidable a problem as it seemed before.

A permanent organization to foster and develop means for accomplishing the stabilization of our globe on its present axis will result when popular support is available. In the meantime, physical research and popular education should be carried on by interested individuals and groups to whom the following research opportunities are suggested:

1. An aviator flying over the Dead Sea reported seeing a submerged city. An organized research effort, based on this report, is herewith suggested. Remains of prehistoric dwellings will be found beneath seas and lakes. Timbers—submerged in water and thus protected from oxidation—will last for ages. Building materials—such as stones and clay bricks—are dissolved by water only very slowly.

Reports and rumors of submerged dwellings, throughout the world, should be gathered and checked for accuracy. The locations should then be visited and explored officially as a basis for gaining scientific information. The human interest aspect of the publicity engendered by such exploration will furnish a broad basis for securing widespread public support for promoting global stabilization.

2. Evidence of glacial action is bound to be discovered in North Africa. The grinding and tracking marks of moving glaciers on rock formations, and the drift of the till, will be found to radiate from the central Sudan Basin. These markings will be hard to find because they will be deeply buried, as at Ur of the Chaldeans and in Crete; but outcroppings should occasionally occur where the top soil has been washed away. Their discovery will prove to be of significance to science.

3. At times (at present about once every fourteen months) the eccentric centrifugal force of the Greenland Ice Cap works in unison with that of the Antarctic Ice Cap—pulling along the same meridian circle. This appears to increase the menace of the next careen of the globe to maximum dimensions. The conjunction of a nutation, or nodding, of the Axis of Spin with this

off-center "throw" might induce the start of the Antarctic Ice Cap toward the Equator of Force.

A tentative, well-reasoned prediction by a world-famous astronomer, as to when such a climax of forces is likely to occur, will excite the interest of the masses and could be of considerable aid in securing popular support for global stabilization.

The portent of a great deluge, caused by the natural mechanical forces of rotation acting on a transient ice mass, should create at once a general desire to try to do something about it. It does not permit of sitting back and hoping that someone else will do something about it for our benefit.

The United States Department of Defense might well be directed to undertake steps to prevent our extinction. An Antarctic Division of the Department of Defense is suggested—in which Army, Navy, and Air Force personnel and equipment will be used to the best advantage imaginable to defend us against the growing menace of the Antarctic Ice Cap.

We will all join in this work once we know that our lives are at stake!

PART TWO

THE CAUSE OF GRAVITATION

> One of the tragedies of life is the murder of a beautiful theory by a brutal gang of facts.
>
> LA ROCHEFOUCAULD

IT HAS LONG been known that the earth rotates on its Axis of Figure. Astronomers throughout the ages have observed this rotation, as well as the periodic diversions from regularity, wobbling and precession, and they have calculated the frequency of revolution to a high degree of mathematical accuracy. But this computation alone does not scientifically account for the rotation of the earth. We must also identify the source of energy that *causes* the earth's rotation. This identification, only recently discovered, also provides an understanding of the nature of gravity and gives scientific validity to the theory that polar ice caps, when they grow oversize, cause successive roll-arounds of the globe.

We find that the cause of the earth's rotation is the all pervading force of nature known as celestial radiant energy. The electrical energy rays from celestial space, shot out by countless billions of stars (suns), collide with the earth, are absorbed by it and create its materials. These celestial energy rays create the phenomena of weight and, striking unevenly, cause the earth to rotate.

This section will undertake to demonstrate that the phenomenon of weight is caused by celestial electrical radiation impinging upon, penetrating, and being absorbed by the earth's materials, and that the cause of terrestrial gravitation is the simple basic phenomenon of dynamic electrical repulsion. Gravitation is to be considered here as an electrical phenomenon!

I

Sir Isaac Newton on Gravitation

THE CURRENTLY prevailing theory of "Universal mutual attraction of masses" will be shown to be:

1. refuted by the observable motions of celestial bodies, falling bodies, and other phenomena
2. based on unproven assumptions
3. unsupported by satisfactory repetitive physical proofs

The Drag of Gravity theory will be presented and supported by the citation of numerous physical evidences supporting gravitational repulsion, and specific evidences of:

1. the existence of celestial electrical rays
2. the forces of rays
3. the penetrating powers of rays
4. the ability of materials to stop rays

Just as our pioneer forefathers had to clear the ground before planting their sustenance crops, so it is necessary to clear away the fallacies of the prevailing theory of attraction of masses, now universally taught in our schools. The statements refuting the theory of attraction of masses and those supporting the Drag of Gravity theory are therefore assembled together.

A venerable and generally respected academic theory states that "bodies attract each other directly in proportion to their masses, and inversely in proportion to the square of the distances apart." This theory of "innate attraction" does not hold up against an abundance of factual evidence to the contrary.

The mass of a body does *not* influence its falling speed. All bodies, large and small, fall to the earth at the same rate of speed. This was first proved by Galileo, and was recently confirmed by scientists at Princeton University. In a vacuum, a feather and a bullet fall at the same speed.

The square of the distance has no relation to the speed of a falling body. The speed of a falling body is *not* increased by a factor of sixteen at one quarter the distance to the earth; it does increase at a fixed rate of approximately 32.2 feet per second during each and every consecutive second of its fall.

The irony of this refutation of "Newton's Law of Gravitation" can best be seen by reconsidering Newton's own statements.

Newton's First Law of Motion states: "A body at rest or motion will continue in a state of rest or motion unless acted upon by some outside force." This Law of Motion would be untrue if bodies possessed inherent powers of attracting other bodies; so-called innate attraction is not an outside force. Hence, this law specifically eliminates any innate gravitational force in material bodies themselves.

In his *Principia* Newton carefully defined the word "attraction" to mean either attraction *or* repulsion. He wrote that he used the word "attraction" to signify any force by which bodies tend toward one another, whatsoever be the cause. "I here use the word attraction in general for any endeavor whatever made by bodies to approach each other; whether that endeavor arise from the action of the bodies themselves as tending mutually to, or agitating each other by spirits emitted; or whether it arises from the action of the ether or of the air or of any medium whatever, whether corporeal or incorporeal, any how impelling bodies placed therein toward each other."

Newton foresaw the misunderstanding that would develop from his simplification of terminology, and specifically attempted to disassociate himself from an oversimplification of his own views: "I desire that you would not attribute innate gravity to me. . . . that gravity should be innate . . . is to me so great an absurdity . . . that no competent thinker can fall into it" (quoted by P.E.B. Jourdain in *Monist*, Vol. 25, 1915, page 252.) And,

Sir Isaac Newton on Gravitation

again, he wrote: "You sometimes speak of gravity as essential and inherent in matter. Pray do not ascribe that notion to me; for the cause of gravity, I do not pretend to know."

These quotations from Newton's letters have been published by others, as well as by Jourdain, who was a professor at Cambridge where Newton also lectured. They need no further validation. But they do bring us face to face with having to determine what is fact and what is fiction in present-day academic concepts of gravitation. Newton evidently judged others by his own competence as a thinker, and it now appears that he was overcomplimentary to many of the savants of the day. "If anyone should explain gravity and all its laws by the action of some subtle medium," Newton wrote, "and should show that the motions of the planets and comets are not disturbed by this matter, I should by no means oppose it." ("Subtle medium" is, today, a synonymous term for defining electrical phenomena.)

Since the time of Newton the theory of the universal mutual attraction of masses has been developed by writers from hypothesis to an ostensible fact. This build-up is exposed by the following excerpts from an article on gravitation by R.S. Ball in the 9th edition of the *Encyclopaedia Britannica*:

"A body dropped from a point above the surface of the earth always falls in a straight line which is directed toward the center of the earth. . . . The observed facts are therefore explained by the *supposition* that the earth possesses a power of attraction." But, in section III he says: "Observations of the most widely different character have combined to show us that this *law*, which was discovered by Sir Isaac Newton, is true. It is called the *Law of Gravitation*." (Editor's stress.)

It is apparent that the validity of the theory of gravitation, which Newton left to others to explain, is, to Professor Ball, at first a "supposition" that mass attracts mass; but later on this supposition becomes the LAW OF GRAVITATION.

In the 11th (1910) and 13th (1926) editions of *The Encyclopaedia Britannica* it is clear that the theory of attraction has progressed further. Here it is stated: "The law of gravitation is unique among the laws of nature, not only in its wide generality, taking the whole universe in its scope, but in the fact that so far as yet known, it is absolutely unmodified by any condition or cause whatever. . . . The general conclusion from everything we see is that a mass of matter in Australia attracts a mass in London precisely as it would if the earth were not interposed between the two masses."

Writers in encyclopedias, dictionaries, and textbooks have credited Newton with the setting up and establishing of what is now called Newton's Laws of Gravitation. Thus the belief in the unproven theory that mass attracts mass, with its four unproven assumptions, has grown until it is now generally taught, and regarded as a fact.

Mass teaching is based on individual learning, upon which it depends for corrections, rationalizations, and revisions.

The four assumptions of the Law of Gravitation are:

1. Mass attracts mass
2. There is a gravitational constant for mass
3. The "constant" is constant for all matter
4. The attraction varies inversely as the square of the distance

The Encyclopaedia Britannica, 13th (1926) edition, under "Gravitation", states: "The law of gravitation states that two masses M_1 and M_2 distant d from each other, are pulled together each with a force GM_1M_2/d^2, where G is a constant for all kinds of matter, the *Gravitation Constant*. The acceleration of M_2 toward M_1 or the force exerted on it by M_1 per unit of mass is therefore GM_1/d^2. Each pulls the other by a force equal to the number of units of mass multiplied by the constant G."

The "gravitation constant G" is not the acceleration of gravity 32.2 feet per second, known to exist at the earth's surface, and referred to in physics as "g". We know that "g" exists with relation to an apple falling to the earth. We know nothing of any force compelling the earth to fall toward the apple, aside from philosophical conjectures.

Physicists have made heroic efforts on the basis of the theory that mass attracts mass. Many experiments have been made during the past one hundred and fifty years in an endeavor to establish a figure for the mean density of the earth. The failure to obtain repetitive duplicate results appears to be the most outstanding fact revealed. The following is condensed from *The Encyclopaedia Britannica*, 1941, Vol. X, page 663, under "Gravitation:"

"The aim of the experiments to be described here may be regarded either as the determination of the mass of the earth in grams M most conveniently expressed by its mass divided by the volume, that is its mean density Δ M or the determination of the gravitational constant G. Corresponding to these two aspects of the problem there are two modes of attack. Suppose that a body of mass m is suspended at the earth's surface where it is pulled with a force w vertically downwards by the earth m its weight. At the same time, let it be pulled with a force p by a measurable mass M, which may be a mountain, or some measurable part of the earth's surface layers, or an artificially-prepared mass brought near m, and let the pull of M be the same as if it were concentrated at a distance d. The earth pull may be regarded as the same as if the earth were all concentrated at the center, distance R.

$$\text{Then } w = G \cdot 4/3 \, \pi \, R^3 \Delta m/R^2 = G \cdot 4/3 \, R \Delta m \quad (1)$$
$$\text{and } p = GMm/d^2 \quad (2)$$
$$\text{By division } \Delta = \frac{3M}{4\pi R d^2} \cdot \frac{w}{p}$$

If, then, we can arrange to observe w/p, we obtain Δ, the mean density of the earth."

Here are the figures for this mean density as arrived at by various experimenters also taken from *The Encyclopaedia Britannica*:

BOUGUER'S EXPERIMENTS

Quito—Isle of Inca:
"Bouguer found the density of the earth was 4.7 times that of the plateau—a result certainly much too large."

Mount Chimborazo:
"He concluded the earth was 13 times as dense as the mountain —a result several times too large . . ."

Maskelyne's Experiments

Schiehallien Mountain:
"Charles Hutton . . . found that the deflection should have been greater . . . arrived at a 'mean density' of the earth at 4.5, a figure later revised by Playfair."

Airy's Experiments

Mean density of the earth finally resolved at 6.565.

Von Sterneck's Experiments

"The values which Von Sterneck obtained for mean density of the earth were not consistent, but increased with the depth of the second station" (depth of the mine).

Cavendish's Torsion Balance and Modifications		Values Obtained for Delta (mean density of the earth)	
Cavendish	1797	5.448	
Reich	1838	5.49	
Bailey	1841	5.6747	
Wilsing	1887	5.79	
Boys	1891	5.527	(only
Braun	1896	5.527	duplicate)
Estavos	1896	5.53	
Burgess	1901	5.55	

Experiments based on determining delta by measurements of decrease in weight with increase in distance (elevation above the earth):

Experimenter	Values for Delta
Von Jolly, 1878-81	5.692
Poynting, 1878	5.493
Richarz and Krigar-Menzel, 1884	5.05

The acid test of a law of nature is that it will, when properly measured, yield repetitive duplicate results. From the above information it is obvious that the theory of attraction of masses has failed to receive the support of indisputable duplicate repetitive measurements.

It is remarkable that this now venerable theory that mass attracts mass has engaged the conscientious efforts of brilliant physicists, who sensed and believed that they had a truth within their grasp. The names of these men have come down to us because they stood foremost among their peers. They always knew that their theory required physical proof, and they labored long, diligently, painstakingly to secure convincing proofs. In part they were limited by their lack of knowledge of radio broadcasting and communications, of radar or of electronics, which are at the recently conquered frontiers of knowledge and which constitute a young, vigorous and revolutionary department of science. (Indeed, the first of the above named experimenters was actually measuring the shielding effects of mountain masses to radiant energy, and not the "pull" of the materials.) Yet, as is often the case in scientific investigation, the tools of discovery were only just becoming available.

It is an interesting coincidence that the theory of the attraction of masses began to be developed at about the time that the attraction and repulsion of electrically charged conducting bodies were being studied by scientists. It may also be a coincidence that some of those scientists thought the electrified conducting bodies attracted and repelled each other, failing to consider the possibility that the bodies did not *themselves* attract and repel, but that they merely hold electrical charges which cause the ether to move them about—as explained elsewhere in this book. ("Static Electricity," page 192). Yet, in the measurements of attraction and repulsion of electrically charged conducting bodies, the same results are always obtainable under the same conditions—repetitive duplicate results which meet the acid test of proof of a true law of nature. This fact, which earlier scientists failed to apply to their concepts of gravity, is an integral part of the theory of gravitational repulsion based on celestial radiant energy.

Indeed, proofs are as lacking for the universal attraction of masses as they are abundant for electrical attractions and repulsions. The force of the assumed gravitational "pull" based on the so-called gravitational constant for all matter, G, has never been measured in dynes of force or pounds pull. The Drag of

Gravity is measured directly in pounds of weight for any material.

By analyses and deductions with observable data, the Drag of Gravity theory has been proved entirely consistent, and the forces of nature, indicated as governing its operations, are now identified. Quantitative measurements are lacking; but these should be produced reasonably soon. Empirical measurements in aid of the advancement of science are natural functional operations of endowed research organizations, of the thesis and postgraduate work of universities, and of industrial laboratories and others possessing the necessary skills and facilities. Time, effort, and the evolutionary development of scientific techniques are the only requirements.

II

Physical Evidence Refuting the Universal Mutual Attraction of Masses Theory

To LIMIT A refutation of the theory of attraction of masses to but scant empirical data would be inadequate to sustain such refutation. But the absence of physical proof is buttressed by the theory's own inconsistencies and theoretical inadequacy to explain observable phenomena.

The first refutation lies in the fact that the earth does not attract meteors, and is based on elementary observation. The records show that meteors strike the earth haphazardly at all angles, while others approach very close to the earth and pass by without hitting it. This clearly indicates that the meteors are not being attracted by the earth according to the theory of mutual attraction. On the other hand, if mass attracts mass directly in proportion to the product of the masses and inversely as the square of the distance between the masses, then the earth would not only attract shooting stars but all meteors would be attracted in virtually a "bee-line" straight toward the center of the earth and would all strike roughly perpendicularly to the earth's surface.

The fact that meteors are not attracted in this manner makes it clear that the mass of the earth does not attract the mass of shooting stars, and that therefore, in this special case, mass does not attract mass. However, due to the shielding effect of the earth in regard to radiant energy coming from the opposite direction, slight deflections of meteors toward the earth are observable—especially as they get near the earth. (See page 170.)

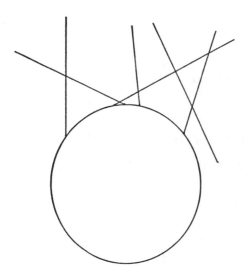

Fig. 7. Meteors strike the earth haphazardly at all angles. They do not come directly toward the center of the earth and do not strike perpendicularly to its surface; this shows that they are not being pulled by the mass of the earth toward its center.

The second refutation of the universal mutual attraction of masses is based on the elementary fact that the acceleration of gravity is roughly a constant, approximately 32.2 feet per second for each and every consecutive second while a body is falling at or near the earth's surface.

If mass attracts mass directly as the product of the masses and inversely as the square of the distance, the acceleration of gravity would not be a constant; but the speed of falling bodies would be accelerated in a geometrical ratio as they approached the earth in response to a constantly increasing "pull" of gravity. The gravitational pulls would, for example, be four times as great at one-half of the distance and nine times as great at one-third of the distance between the falling body and the earth.

The speed of the falling body would respond to the intensity of the pulling force—just as the initial speed of a bullet responds to the intensity of the driving force. It is clear that a different driving force produces a different speed. It is equally clear that a geometrically increasing pulling force must produce a geometrically increasing acceleration of gravity, and that this must occur automatically with decreases in distance. If this is not a physical

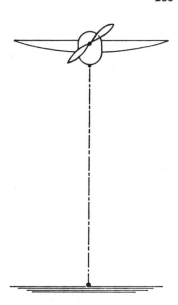

FIG. 8. The acceleration of gravity is a constant. The speed of a bomb dropped by an airplane increases at the steady rate of 32.2 feet per second for each second of its fall. The speed of its fall is not increased faster than the steady rate of increment as it nears the earth. There is therefore no such thing as a pulling power varying inversely as the squares of the distances between earth and bomb. The earth does not "grab," like a magnet, as distances are decreased.

reality, the theory that every particle in the universe attracts every other particle, and that the pull is inversely as the square of the distance, is proved to be untrue. Hence, because the acceleration of gravity is a constant, it is a fact, proved by tests of many observations, that mass does not attract mass inversely as the square of the distance.

The third refutation of the universal mutual attraction of masses and of the theory that the attraction increases directly as the product of the masses, is based on empirical tests of falling bodies. Galileo, in the seventeenth century, dropped weights from the Leaning Tower of Pisa and announced the discovery that a one-pound weight fell to the earth at the same speed as a ten-pound weight of the same material. His experiments proved that a large mass did not fall faster than a small mass. This was a proof that the so-called "attraction" did not vary directly as the mass. Others have made similar experiments that support Galileo's tests under different circumstances, and with greater mathematical accuracy. In ballistics a cannon ball and a rifle bullet follow the same trajectories, with modifications for air resistance, showing that both fall to the earth at the same rates of speed.

In the case of bodies falling to the earth, the facts known to science, therefore, refute the theory of attraction of masses directly as the product of the masses and inversely as the square of the distance between them!

A currently accepted explanation under the universal mutual attraction of masses theory, that helps to account for constancy of constant G near the earth's surface, says that we must consider the gravitational "pull" of the earth mass as if it were all concentrated at the center of the earth. This explanation seems to be an additional theory invented to support the original erroneous theory; since it denies the attraction of mass for mass on the surface of the earth, the result is a denial of the theory of the universal mutual attraction of masses.

If gravitation were an attraction between masses, then there would be a difference in weights of bodies on the surface of the earth at noon and midnight, for the reason that the centrifugal force of the earth, due to the rotation of the earth on its axis, is tending to throw materials toward the sun at noon and away from the sun at midnight. This centrifugal force exists. It requires a correction of ship's clock pendulums when moving toward or away from the Equator—which is equivalent to stating that a given piece of material actually weighs less at the Equator than at the North Pole or the South Pole. Further, if the assumed attraction of the sun existed, then its force—combined with the existing centrifugal force of earth materials—would show less weight at noon, with the sun overhead, and a greater weight at midnight for any test piece. Yet no decreases in weights of materials due to the assumed "pull" of either sun or moon, have ever been disclosed to exist. The facts once more indicate that no attraction exists between objects on earth and the sun or moon.

The currently prevailing explanation of why the heavenly bodies do not coalesce, due to the heretofore assumed mutual attraction of masses, is that the "original forces" act to keep the celestial bodies moving in a straight line and that their constant fall towards each other is counteracted by the inertia due to these "original forces."

Refuting the Masses Theory

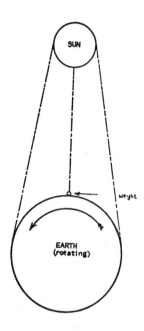

FIG. 9. The sun exerts no "pull" on a test weight. At noon, with the sun directly overhead, any test weight would weigh less if the sun is pulling on it, while the centrifugal force, due to the earth's rotation, is tending to throw it toward the sun. No such decrease in weight has been discovered.

The theory of "original forces" (set in motion at the time of creation) is proved erroneous by the extreme variations in the speed of the earth in its orbit around the sun and by the variations in the speed of the moon in its orbit around the earth.

Motions due to inertia cannot change their speeds, except to die down, while the speeds of the earth and of the moon, in their orbits, are variables, becoming faster and slower at regular recurrent intervals.

If the velocity of a body be increased, the force producing that velocity must have been increased, or the resistance to its motion decreased. And if the velocity of a body be reduced, the impelling force must have been reduced, or the resistance increased.

The theory of "original forces" is, therefore, without scientific foundation. The alleged forces simply do not exist. The introduction of the idea of such a force in the attempt to bolster up the theory that mass attracts mass is thus seen to be like a boomerang

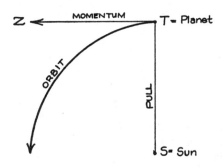

Fig. 10. Old theory of why the heavenly bodies do not crash together. The curved line is the orbit of the planet around the sun, resulting from the combined actions of force TZ and force TS. The line TZ represents the "original momentum." The line TS represents the assumed "pull of the sun." Neither of these forces exists.

which returns and destroys the original theory of attraction of masses when the theory of "original forces" is exploded.

A corollary to the refutation of this long-held theory is that if masses—such as sun, earth and moon—are attracting each other, and there is no "original force" of momentum to hold them apart, they would come together quickly. The facts are that they do not come together—hence, there is clearly no universal mutual attraction of masses.

As we saw above, *The Encyclopaedia Britannica* has, in its past editions, given authoritative expression to developments in scientific thought up to the time of publication.

Significantly, however, a change of attitude toward "attraction of masses" is demonstrated by the successive writers.

It is proper to add that the 11th (1910) edition refers only very briefly—and by abstract mathematical equations only—to the acceleration of planets toward the sun, and of the moon toward the earth.

Both the 13th (1926) edition, and the 1941 printing, skip entirely these items of mutual planetary attraction, and make no attempt to explain away the failure of the sun and earth and moon to crash together due to the law of universal mutual attraction of masses. This suggests a trend of scientific thought away from the older theories, but does not indicate what new theory is to supplant the old.

III

THE GROWTH OF THE THEORY OF GRAVITATION REPULSION

As WE HAVE seen, Sir Isaac Newton was strictly neutral as to whether gravitation involved a pull or a push. He was careful to explain that he made use of the word "attraction" between masses to mean either attraction or repulsion, or any means whatsoever impelling bodies toward each other. He wrote that he would not oppose an explanation of gravitation which laid the cause to some subtle medium, provided it did not contradict astronomical observations.

As suggested earlier, the explanation offered here is that electrical energy is the subtle medium whose identification Sir Isaac said he would not oppose as being the cause of gravitation. This energy is found to occur as both dynamic and static electricity. In its dynamic form, known as radiant energy, it is being broadcast from celestial bodies, causing gravitation and the weight of materials. In its static form it holds the heavenly bodies apart, preventing cosmic collisions, since all celestial bodies possess the same kind of electrical charges on their surfaces and thus repel each other rather than collide, as they would do if they were all attracting each other by some inherent force or pull.

The theory of gravitation by radiant energy is not new. The earliest reference is that made by LeSage, of Geneva, who explained that "ultramundane corpuscles" in space impinged on the bodies in space and caused gravitation. (*Journal des Savants*, 1764)

The Dutch physicist Lorentz (1853–1928) has stated that "The theory of LeSage can be saved by assuming the corpuscles are wholly or partially absorbed by matter, but then the picture is deprived of its simplicity." (*Lectures on Theoretical Physics*, by Hendrik A. Lorentz, translated into English, Vol. I, page 153, Macmillan Co., 1927). This appears to be a prophetic forecast of what is now disclosed as the absorption of energy in creating earth materials and as the cause of gravitation and the Drag of Gravity.

In his book *The Copernican Revolution*, Thomas S. Kuhn explains the genesis of the prevailing theory of mutual attraction of masses. On pages 252-62 he states: "The French scientist Descartes (1596-1650) said, 'loose bodies are driven to the earth by the impacts of aerial corpuscles in the earth centered vortex'". That became an accepted belief at that time. Newton agreed with Descartes (Newton, *Optiks*, 4th edition, 1730, N.Y., Dover, 1952, page 401). "Again and again Newton insisted that gravity was not innate in matter. In spite of his specific intention Newton led most of his successors to believe that gravity, and therefore weights, were intrinsic properties of matter." Cotes, who translated Newton's work from the original Latin, did more than anyone else to establish the present concept of mutual attraction of masses. "It was forty years (of battling) before Newtonian physics firmly supplanted Cartesian physics. Though forced at last to admit his defeat, he (Newton) continued to maintain that someone else would succeed."

Charles F. Johnson, of Trinity College, has published a pamphlet entitled "Gravity Not an Attraction" (1925), in which he states: "The assumption or hypothesis is that all space is full of a force acting in straight lines in every direction on matter. Like magnetic attraction, it is not perceptible to our senses, but, unlike magnetic attraction, it is not increased at short distances. This force penetrates all material bodies. Unlike light, nothing is opaque to it and nothing is transparent. In passing through matter, it parts with a proportion of its energy proportional to the mass or density. It is of the nature of pressure, but unlike the pressure of an elastic fluid, it is not reflected and acts in straight lines only. It would force a yielding mass into a spherical form. It strikes the earth on all sides equally, but in penetrating the earth it loses with every foot a part of its energy.... It beats down on the surface of the earth with millions of pounds to the square foot—sufficient to amount on the entire surface to the weight of the earth.... But the energy comes from an illimitable source—space. The universe being regarded primarily as a universe of force and not of matter."

Professor Johnson made the first mechanical device for detecting variations in the quantity of electrical radiant energy from space, at a given location, by measuring the relative amounts of incoming rays being shut off on the side toward a mountain as compared to the side toward the sea.

In *The Ether Stream* (1921), J. S. Miller of England states: "Ponderable substances are known through our senses. They can be weighed and measured. Imponderable substances are known to us only through their action on ponderable substances. They have no weight. The ether is chief [imponderable].... If imponderable substances can act on and move ponderable substances, then ponderable substances must react to imponderable substances in order to maintain the natural law of action equalling reaction, or equilibrium.... A stone, then, falls to the earth because the ether stream is rushing through it at a great velocity on its way into the earth, and by its frictional force carries the stone to the earth with it.... The stone has no reciprocal power to draw the earth to itself...."

A paper entitled "A Kinetic Theory of Gravitation" was read before the American Association for the Advancement of Science by Charles F. Brush in December 1910. He stated, "I believe that kinetic energy of the ether is the fundamental cause of gravitation." He quoted Sir J. J. Thompson as saying, "All kinetic energy is energy of the ether", and he quoted Sir Oliver Lodge as saying "All potential energy exists in the ether."

He continued, "As is well known, the ether waves of light will exert a slight pressure on a body."

He called to mind a large chamber with uniformly lighted walls, in which two opaque bodies are suspended, and stated, "Each body will be partially shielded by the other body from the ether waves coming from that direction. Hence the light pressures will be less on that side of each body which faces toward the other than on the side which is turned away".

He differentiated between light waves which cause heating and the longer ether waves that do not excite molecular vibrations; and he explained gravitation as a push rather than a pull.

Dr. Brush delivered a lecture on gravitation, in April 1929, before the American Philosophical Society of Philadelphia (Vol. LXVIII, 1929), in which he gave the results of experiments confirming his statement: "Conversion into heat of some of the energy of gravitation ether-waves, however little, might be expected to impair to some extent the falling velocity of a heat generation substance; and all such substances thus far tested have shown impairment."

He reported and analyzed tests made by him demonstrating that the speeds of acceleration of falling bodies, at the same time and place, were not identical between heat absorbing and non-heat absorbing bodies. When some of the energy which caused the speeds of falling, was absorbed by the body itself, then the body fell more slowly.

The above statements were referred to the Gravity Research Foundation of New Boston, New Hampshire, and in reply the president, George M. Rideout, has written me that, "Dr. Brush's ideas regarding the gravity effect of heated bodies has been shown by Poynting and Phillips to be less than two parts per million per degree. That is less than the limits of experimental error. Consequently there is no verifiable evidence for Dr. Brush's Theory."

The above comment states, in effect, that there might be a recognized difference in the speeds of falling bodies of heat absorbing and non-heat absorbing materials, but that the difference is too small to detect experimentally. A communication from the National Bureau of Standards states that the effect Dr. Brush proposed has never been verified.

"The Analogy between LeSage's Theory of Gravitation and The Repulsion of Light" by G. H. Darwin, F.R.S., was read on May 18, 1905 and is recorded in the *Proceedings of the Royal Society of London*, 1905. He treated the subject mathematically and referred to contemporaneous works on repulsion by Poynting and Lord Kelvin.

IV

DYNAMIC ELECTRICITY

LOOKING OUT INTO SPACE we see billions of suns radiating energy. We find that we are in a universe of force and that the earth on which we live is a ball of matter and minor electrical forces reacting to a surrounding universe of electrical forces. It has been estimated that the total radiation into space from the sun alone amounts to 3.79×10^{33} ergs (of energy) per second in the form of light, heat, and other ethereal vibrations, and that only one part in 120 million hits a planet or star directly.

If we point a telescope into the sky and focus its eyepiece on a selenium cell, or phototube, this "electric eye" will detect and respond to the incoming celestial radiant energy. For the spectacular openings of the Chicago Century of Progress Exposition a phototube was used, while for the opening of the New York Sesquicentennial Celebration a selenium cell was used. In both cases the telescopes were pointed at the star Arcturus, for sentimental reasons only, and the cells reacted after a fraction of a second's exposure to the incoming energy rays.

If we attempt to measure in this way the total incoming energy, only guidelines can be given here. The known factors are sketchy and will require much further research. We know the minimum amount of energy required to operate the photoelectric cell; we do not know what maximum input could be, and our knowledge of the efficiency of the operation of the cell, a factor in the computation of the total amount of received energy, similar to the factor of transparency of the lenses to the wave lengths involved, is limited to our ability to compare the photoelectric cell to other standards. We know that only a fraction of a second was required to operate the cell; but in our computation we must adjust this figure to the energy received per second, all the while taking into account the specific physical properties of the photocell that manifest themselves in the change from an initial reaction to a pulse of energy and a continued exposure to this energy.

With a known angular field of telescopic lenses, and with a knowledge of the factors just mentioned, it is a matter of simple arithmetic to calculate the total energy that impinges upon the earth per second as a result of the energy vectors received by the telescopic field. By taking various readings of the photoelectric cell at various angles of the telescope to the horizon, a reasonably accurate approximation of the total energy impinging upon the earth from all angles can be developed mathematically.

The minimum amount of radiant energy of wave length 0.8 microns detectable by a selenium cell is about 10^{-12} watts. Knowing this factor, which is the minimum amount of energy coming through the telescopic lens to make the tube work, and assuming that the average lengths of the radiant energy rays from celestial space is 0.8 microns and the exposure a full second, we find the total calculated incoming daily energy from celestial space to be approximately 2¼ billion B.T.U.'s (British Thermal Units), which is the equivalent of about 2 trillion foot pounds. (The data on energy required to make a selenium cell function were furnished by the National Bureau of Standards.)

The above is an approximation, because of the unknown time element involved. It indicates that the energy from celestial space, calculated as above, is probably a tiny fraction of the total energy received daily by the earth. Waldemar Kaempffert quotes the work of Dr. Charles G. Abbott of the Smithsonian Institution and states that "In the deluge of sunshine that inundates the earth every hour there is the energy equivalent of 21 billion tons of coal." (*Explorations in Science*, Chapter 14). Basing our estimate on coal having 14,000 B.T.U.'s per pound, we arrive at an astronomical figure of B.T.U.'s per day: 14,000,000,000,000,000,000.

Dynamic Electricity 179

The relative amounts of electricity received by the earth from the down-pouring celestial electrical radiations which cause materials to have weights, and the amounts of electricity received directly from the sun are yet to be evaluated. Light and heat on the surface of this planet are dependent upon the sun. Animal and plant life have heretofore been considered as responding to the sun alone; but, animals and plants grow at night when direct sunlight is absent. Corn is known to grow faster at night. The night blooming cereus opens only at night.

Energies in motion in the form of electrical radiations are forces, and energy rays are corpuscular. Electrical energy radiation from celestial space has photographed itself by striking atoms in the emulsions of photographic films. The pictures produced show the streaks of the incoming cosmic ray particles, and also the streaks of the splashed-out fragments, appearing something like a high-powered bullet striking a bag of marbles which become scattered by the impact. The streaks caused by the fragments of the smashed atoms indicate that they become secondary forces in motion before becoming dissipated in the earth or its atmosphere.

The use of the analogy of bullets requires qualification, because the photons, quanta, and the cosmic rays and products are continuously penetrating downward through all animals, without their becoming aware of the existence of the incoming energy radiations. Animals are similarly unconscious of the radio and television radiations which also permeate the ambient atmosphere and pass through them.

The equivalence and interchangeability of matter and energy are clearly indicated by the spectroscopic analysis of materials. When heated to incandescence, all materials broadcast the unique energy rays of their constituent elements in definite segments of the spectrum, corresponding to the energy levels of the electrons within the atomic structure. Thus, by the use of a standard table of the constant patterns of emission of each element, the chemical structure of any earth material can be positively identified.

This technique indicates that all earth materials are built up of dynamic radiant energy rays which were metamorphosed to concrete materials by animal, vegetable, and mineral growth—a theory fully investigated in Part Three. The change back from materials to radiant energy occurs on heating to incandescence, whereby the materials lose their molecular identities through disintegration and reassociation into gaseous molecules that in turn radiate characteristic dynamic energy rays.

Proofs of the interchangeability of energy and matter are the atomic bombs, in which matter changes into energy instantly, and photosynthesis, where energy is slowly converted into the plant materials of the earth.

Because dynamic electrical rays are imponderable substances they can be identified only by the effects they produce on ponderable substances. The oneness of energy, matter, and dynamic electrical radiations is indicated by the "Edison effect"—a term used because of his early research and identification of the cause of the lighter-colored shadows or lines on the insides of clear glass incandescent light bulbs of the long carbon filament type, in which one leg of the filament intercepted the flow of radiant energy from the other leg.

When energy rays are thus produced and radiated by the incandescent filament in a glass electric light bulb, the existence of a force or energy being emitted is evident. Edison proposed signalling to ships at sea by means of this source of energy power.

Particles, as well as light and heat rays of the incandescent materials, are shot away, and some of these emissions, which are not unlike microscopic shooting stars, are stopped—or metamorphosed from energy back to matter—and appear on the inside surfaces of the glass bulbs through which the light rays pass.

The materials which appear and build up on the insides of lamp bulbs, when incandescent, were once a part of the filament of the lamps; we thus have the phenomenon of materials changing from matter to radiant energy, or, like meteor dust, passing across a short space from filament to glass, in a vacuum, and a very small part of it reappearing as matter on the insides of the bulbs, when stopped by the glass.

Action and reaction being always equal and opposite in direction, it follows that the glass was subjected to a tiny element

Dynamic Electricity

of dynamic electrical repulsion, which is something analogous to the dynamic electrical repulsion of the radiant energy rays and meteor dust being continuously poured from incandescent spheres in celestial space upon all elements of the surface of the earth.

As suggested earlier, it is this dynamic electrical energy of celestial rays that is responsible for terrestrial gravitation. Weight is caused by the downward impact and penetration of celestial electrical radiation releasing its energy in the materials with which it collides, and by which it is retarded and eventually stopped.

The materials of the earth are composed of molecules, which in turn are composed of atoms. Atoms differ in their number of positively charged protons, neutral neutrons, and negatively charged electrons. The more each atom possesses of each of these atomic particles, the greater its chances of being struck by incoming energy rays, and hence the heavier its weight. The permeability of atoms of the earth materials to the incoming energy radiations from celestial space determines weight. Weight varies inversely with permeability.

Atoms can be represented as individual entities, resembling the units of our solar system and consisting of a nucleus of protons, charged with positive electricity, with an equal or greater number of neutrons, without electrical charge (except hydrogen), and a number of electrons, equal to the number of the protons and charged with negative electricity, whirling about the nucleus like the planets that revolve around the sun at definite but different distances.

The revolving electrons of the heavy earth elements are assumed to occupy definite zones and revolve in circular orbits about the nucleus. They are not all concentrated along the same orbit around the nucleus. Thus, the heavy element uranium has 92 protons and 146 neutrons in its nucleus, and 92 electrons which are whirling about the nucleus, grouped in definite zones at seven different distances from the nucleus.

The radii of the whirling electrons of the atoms are extremely great in extent compared to the relatively tiny sizes of the incoming radiant *energy*, hence the chances of a collision with the incoming energy radiations are extremely rare; but every energy atom from celestial space eventually strikes a nucleus or electron of an atom of earth material. Earth elements, such as uranium, which possesses many protons, neutrons, and electrons, are struck by the incoming rain of energy atoms, much more often than earth elements with fewer such particles; therefore uranium is heavier than an element such as hydrogen, which has only one proton, one electron, and no neutrons.

If the radii are short, or if the atoms are close together, there are more chances of collisions and the atoms will be struck more often by the incoming energy radiations than if the radii were longer and the spaces between the particles greater.

Each different kind of material produces a different stopping power for the electrical radiations, and this stopping power we recognize as its weight. The effect of these forces, coming from all angles and penetrating all material objects near the earth's surface, is to tend to *drag* or carry the materials along in the directions of their flows.

The sum total of the actions of all the forces impinging on and penetrating materials and of the reactions of the materials to these forces—together with radiations from the earth into space—results in a downward component nearly perpendicular to the surface of the earth, or as the surface would be if all mountains and valleys were reduced to a common plane—action and reaction being equal and opposite in direction.

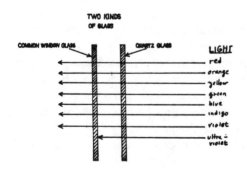

Fig. 11. Common window glass stops the passage of the ultraviolet rays, but permits passage of the light rays. Quartz glass permits the passage of the ultraviolet rays.

Dynamic Electricity

Energy rays of many different wave lengths or frequencies reach the earth from celestial space. They differ in their effects on earth materials. Window glass stops the ultraviolet light rays but permits all other light rays to pass; this proves that the stopping power of each material varies with the different wave lengths of electrical radiations.

Quartz glass permits the passage of ultraviolet rays but common window glass stops the very same rays; this proves that there is a variation in the stopping powers of different kinds of materials when exposed to electrical radiations of the same wave lengths.

These and similar examples demonstrate conclusively that light rays of different wave lengths possess different penetrating powers, and that a material can stop the light rays of certain wave lengths but will permit energy rays of other wave lengths to pass through it.

These elementary citations indicate fundamentally different and varying reactions between earth materials and the incoming radiant energy rays from celestial space.

Dynamic rays, like bullets, are absorbed through resistances similar to friction, but, unlike bullets, the rays cause no apparent disruption of the materials penetrated. Not the bullet itself, but its force is absorbed by the reactions of materials to its passage through them.

Radiant energy rays are forces in motion; they not only possess force but are synonymous with force; and, because energy is indestructible, these radiant forces are not lost—whether emitted from spheres in the celestial heavens or from the materials of the earth.

A radiant energy motor, or radiometer, which demonstrates repulsion is widely sold as a scientific toy. It consists of a small clear glass globe—somewhat resembling a large-size incandescent lamp bulb—in which a vacuum has been created. Four flat vanes are mounted vertically and spaced like the arms of a horizontal windmill, supported on a needle pivoted shaft. One side of each vane is coated black; the other side is white. Black absorbs light and heat radiations; white does not. The result is that the black sides of the vanes are repelled by the light and heat radiations, which they absorb. The vanes, therefore, revolve on the shaft at speeds proportional to the intensities of the light and heat rays—producing a simulation of perpetual motion.

Light rays are today recognized as a medium of energy transmission, similar to an electric current in a circuit, and are being used in the functioning of photoelectric tubes, or "electric eyes," just as microwaves of another wave length are used to transmit communications.

The gamma ray, for example, has a penetrating force like that of a bullet. This force may be measured by the stopping powers of materials; therefore, because action and reaction are always equal and opposite in direction, the force of the gamma ray, its dynamic electrical propulsion, is balanced by and measured by the stopping powers of materials.

The gamma ray penetrates 1) 22 feet through water, and 2) 2 feet through lead, as reported by Robert A. Millikan, the well-known research physicist. The specific gravities of water and lead are 1 and 11 respectively; these are the same proportions, but in inverse ratio, that we noted regarding the penetrating power of the gamma ray.

This illustration proves 1) the power of materials to stop radiant energy rays, and 2) the absorption of rays by materials which limit the penetrating power of the gamma rays. It associates the power of materials to stop gamma rays with the Drag of Gravity by disclosing the same relative drag by water and lead in the case of gamma rays as well as those incoming rays from celestial space which cause the phenomenon of the weight of materials.

Some of the light rays from the sun penetrate materials while some are stopped by materials. They penetrate water and glass, but are stopped by wood and stone; but if the water is deep enough, or the glass thick enough, the light rays are finally stopped and absorbed.

Experiments by Fizeau and Foucault showed that light rays are slowed down by water, indicating a reaction of the water equal and opposite to the dynamic forces of the energy rays.

Dynamic Electricity

Rays from the sun strike our skin and cause sunburn and tan. They are also reported to be able to penetrate our bodies; that light and heat rays can penetrate opaque bodies is indicated by the fact that our internal organs respond to sun baths and toughen at the same time that the skin on the surface of our bodies become tanned.

The theory that celestial electrical radiations impinge on and penetrate matter, causing a downward drag or *repulsion* of gravity, also assumes that the forces producing the downward drag of gravity are fairly uniform and steady and will cause the acceleration of gravity to follow an arithmetical progression of approximately 32.2 feet per second for each and every consecutive second during which a freely falling body is falling. Observation supports this theory and positively refutes the prevailing theory of a geometrical progression.

The phenomenon known as weights of materials is produced by the impact of these forces as they collide with and are absorbed by the earth. The relative capacity of each type of material to stop celestial electrical radiations is measured directly in pounds and ounces.

The earth, by stopping and absorbing these energy rays, becomes a shield against the forces coming from the direction of the antipodes. The downward forces predominate at any point on the surface of the earth.

The energy of these bombarding rays is not lost. It is absorbed in the earth by the materials of the earth. Some of the rays are creators of materials, as outlined in Part Three. The impacts of various rays are not absorbed to the same degree by all materials, but are absorbed by each material in proportion to a phenomenon called weights of materials.

Weight measures the power of materials to stop radiations from space. The dynamic energies of the bombarding celestial radiation are converted into weight, by virtue of which the materials are forced toward the approximate center of the earth.

Physical phenomena to be subsequently identified and analyzed show a slight west to east predominance of the incoming radiant energy rays, a fact which causes the globe to rotate from west to east. The position in space of the earth's Axis of Spin is established by these static and dynamic forces, just as its Axis of Figure is established by the materials of the earth.

The effect of these forces, coming from all angles and penetrating all material objects near the earth's surface, is to tend to drag or carry the materials along in the direction of their flow. The sum total of the actions of all these forces, impinging on and penetrating materials, and of the reactions of the materials to these forces, together with radiations from the earth into space, results in a downward component perpendicular to the surface of the earth—action and reaction being equal and opposite in direction.

Today *atoms* are being smashed into *atoms* by the various high voltage cyclotrons.

Weightlessness is a phenomenon which occurs in space because the celestial energy radiations strike the atoms of the various materials from all directions and therefore neutralize each other; however, at the surface of the earth the impact of the radiations exerts a uniform pressure directed toward the center of the earth, the reason being that the earth itself is interposed as a shield against radiations from the antipodes.

Since one needs physical proof in order to establish the validity of any new theory of science, a review of various examples of recorded and observable evidence follows.

If we could look down from the North Star at the plane of the earth's orbit, with the sun in its center, we will see that the earth rotates counterclockwise. If it did not rotate as it moves forward along its orbit, it would continually keep a section of its forward left side away from the sun's rays, while fully exposing a similar section of its rear right side to those radiations.

In this case if bullets were being shot from sun to earth, instead of rays, more bullets would hit on the exposed rear right areas and fewer on the forward left areas that were constantly hidden from the sun's radiations. Such a situation would aid in producing the existing counterclockwise rotation of the earth and supports the theory of dynamic repulsion by radiant energy, this being a part of the Drag of Gravity theory.

Dynamic Electricity

FIG. 12. Assuming that the earth did not rotate!

The orbit of the earth and successive positions of the earth in it are shown, illustrating that the impinging radiations from the sun would tend to force rotation in counterclockwise direction. The earth's rotation *is* counterclockwise.

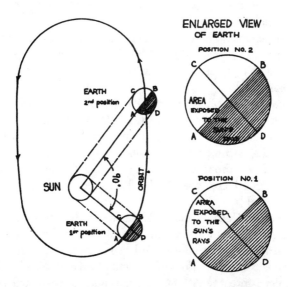

In first position the area of the earth ACB is exposed to the repellant rays of the sun, and the area ADB is shielded.

In second position the area C to B which was formerly exposed is now shielded from the direct rays of the sun, while the area A to D which was formerly shielded, is now directly exposed to the push of the repellant rays.

Imponderable forces are pushing everything ponderable from west to east. They force the earth to rotate from west to east. They cause the ocean waters, the suspended detritus in rivers, the high-low barometric pressures, the winds, the atmosphere, thunder storms and freely falling bodies to travel from west to east.

In the study of ocean currents made by "bottle-casters" and others it has been found that bottles and other freely floating objects, where not deflected by local winds or currents, drift from west to east. It seems to be obvious that this effect is caused by the west to east push of the incoming rays of radiant energy—of dynamic electrical repulsion.

Ocean currents show the effects of the west to east repulsion of radiant energy radiations. The Gulf Stream in the Atlantic Ocean and the Japan Current in the Pacific Ocean are both deflected eastward.

Moreover, the waters of the Pacific Ocean are being pushed against the west coasts of North and South America. The mean ocean level is higher by eight inches on the west coast of Panama, at the entrance to the canal, than on the east coast. In the area of the proposed Nicaragua canal the ocean waters on the Pacific coast are three and a half feet higher than the east coast sea level.

The Pacific Ocean finds relief outlets to the south in the Cape Horn current, flowing from west to east, and to the north through Bering Strait, flowing continuously into the Arctic Ocean.

There is a complex flow of currents in the Strait of Gibraltar, which opens into the Atlantic Ocean, and its eastward flow appears to reflect the west to east push of the Drag of Gravity. "In addition to the normal tidal flow at a maximum of 1½ knots extending from surface to bottom, running both ways at 6-hour intervals, there is superimposed an upper current of about 1¼ knots, flowing from west to east into the Mediterranean Sea, and a west-going lower current of slightly lower speed flowing from the east into the Atlantic Ocean." (Communication from the British Admiralty.)

High and low barometric pressure areas travel continually around the globe from west to east, at fairly well-defined rates of speed, except where local conditions change their speeds and directions. This observable fact requires an explanation, and the rational explanation is the west to east push of radiant energy.

Dynamic Electricity

It is known that areas of low barometric pressure concentrate more moisture than areas of high barometric pressure. In such areas, incoming celestial radiations thus have something more to push against. Naturally, therefore, the low areas travel faster than the high.

It is known that both low and high areas travel faster in winter than in summer. This follows naturally, because the particles—against which the celestial radiations strike—are denser in winter since they are colder, and they therefore receive a greater push from west to east.

The speeds of Lows and Highs in miles per hour in the United States, based on data from the New York City station of the U.S. Weather Bureau, are as follows:

	For entire year	For 3 winter months	For 3 summer months	Per cent increase in winter
Average speed of Lows	28.6 mph	34.8 mph	25.4 mph	42%
Average speed of Highs	25.6 mph	28.4 mph	22.7 mph	25%

Lows are rotated counterclockwise in the northern hemisphere and clockwise in the southern hemisphere, the reason being that the incoming west-to-east radiant energy rays impinge on the moisture particles concentrated in the Lows at more favorable angles on the side toward the Equator, due to the spherical shape of the earth. The Highs then rotate in opposite directions in re-establishing normal pressures.

The positions of the stars overhead may affect the latitudes at which the Lows and Highs travel from west to east; they thus may have an effect on our local weather. As the earth travels through space, slight variations in the intensities of the radiant energy pouring in from celestial space results from the relative positions of the stars above and may cause the Lows and Highs to travel at lower or higher latitudes.

The records of ancient history show that men have instinctively looked to the stars and their positions in the heavens as exerting some effect on global life.

Large masses of summer clouds—known as "thunderheads"—

grow bigger while moving slowly in a sideways direction. But once they begin to precipitate rain, they move from west to east at fairly well-defined rates of speed. This movement of freely suspended collections of electrically bound moisture particles, which the clouds are precipitating as rain, is caused by the west-to-east push of the incoming radiant energy rays.

The United States Weather Bureau advises: "In general, in middle latitudes (in both hemispheres), thunderstorms usually move from a westerly direction. The direction of movement is usually determined by the larger scale flow, in which the thunderstorm cell is imbedded."

It is well known that there is a "Great West Wind" encircling the earth at a high altitude. Dust from volcanoes has been known to go round the world, from west to east, the driving force being the incoming rays of radiant energy.

Incendiary balloons—launched by the Japanese during World War II—brought these high-altitude west winds to public attention. The balloons had control devices to keep them flying at altitudes of 30,000 to 35,000 feet, at which they would drift toward America in the west-to-east air currents; the speeds of these air currents have since been clocked by radar-equipped balloons at 450 miles per hour.

The Mississippi and Rhone Rivers both flow southward. Their deltas are deflected to the east because of the west-to-east push of incoming radiant energy.

Rivers in the northern hemisphere have been found to erode their west banks. They have a tendency to fill toward the east banks, due to suspended detritus being pushed eastward by the west-to-east radiant energy rays and the westerly winds which they induce. The Hudson River at New York City, for example, requires dredging toward the Manhattan side more often than toward the New Jersey side.

The wobble of the earth is caused by the same forces that cause its rotation. The earth is being continually assaulted by photons, quanta, cosmic rays, and meteor dust. Since there is a greater quantity coming from the west than from the east, there is a west-to-east rotation of the globe. As the earth rushes through

Dynamic Electricity 191

space, being a part of the whole solar system, the angles at which these radiant energy rays strike the earth vary by minuscule amounts and bring about changes in the position in space of the Axis of Spin, or the true axis of the earth. The gyroscopic effect of the centrifugal force of the stabilizing bulge of the earth causes the Axis of Figure to adjust itself to the changed position of the true axis—and is the cause of the wobble.

A theoretical question—"Do falling bodies move southward?" —has been the subject of numerous experiments. The theory is that the earth, being bulged at the Equator and also because of the assumed universal mutual attraction of masses theory should offer greater "pull" in a southerly direction on bodies falling freely in the northern hemisphere. One phenomenon noted in all the reports that I have examined is to the effect that an eastward deflection of the falling bodies has always been found. This eastward deflection could be produced by either or both of two causes: 1) The momentum imparted to the body by the faster speed when at high altitude, the object then being farther from the center of the rotating earth, or 2) It could be caused by the west-to-east push on the body by incoming radiant energy rays.

The southward deflection is naturally accounted for by the Drag of Gravity: the shielding effect against the radiant energy coming from the direction of the bulge of the earth, to the south, is sizable compared to the lack of shielding against the energy radiations coming from the north.

V

STATIC ELECTRICITY

STATIC ELECTRICITY differs from dynamic electricity in the manifestations or ways in which it makes its presence known to us. It is the invisible force which keeps the heavenly bodies apart, prevents their collisions, and causes the tides of the oceans.

Static electrical repulsion is one of the forces of nature discovered by experiments on the surface of the earth. Like dynamic celestial radiation, static electricity is one of the prime underlying forces controlling and determining the phenomena commonly attributed to innate attraction. Whereas gravitation has been erroneously construed as an inherent property of masses, it is now explained accurately as resulting from the forces of celestial static and dynamic radiations, and from the interplay of dynamic radiation on materials within the environmental forces and the positions of bodies in space determined by the forces of static electricity.

On a cosmic level, the earth itself is a conducting sphere which is being continuously charged electrically. The relative amounts of the electric charge vary, causing variations in the earth currents, which at times and places interrupt grounded telephonic and telegraphic communications. Variations in earth currents are caused by changes in electric potentials, clearly coming from without and not from within the earth.

The earth is continuously receiving electrical charges in the form of celestial radiations from untold billions of incandescent spheres. It is at the same time discharging electricity into space.

Static Electricity

This system is balanced so nicely that, at its rate of discharge, it would be completely discharged in 7½ minutes, if it were not being constantly recharged from without. (See *Physics of the Air*, by W.J. Humphreys.) The sum total of the outgoing conduction currents, continually flowing away from the earth, like a cataract always falling but never running dry, is balanced by the incoming flow of celestial electrical radiations, which always leaves the earth with approximately the same average charge.

Electricity flows only in closed circuits from points of higher to lower potentials. All that is lost in any circuit is the potential. The flow of electrical energy from celestial space to the earth, and from the earth into space, is necessarily in closed circuits.

It has been known for a long time that there is a steady flow of electricity from the earth into space. There has been a recognized gap in the science of physics, a gap which requires an explanation of how the electricity arrives in the earth. The theory of a celestial radiant energy that produces gravitation, rotation, and build-up of the earth, now fills that gap. It supplies the missing element required to complete the theory of an electrically radiating earth.

This theory requires not only recognition of the incoming electrical radiations as a fact but it also accounts for the long recognized outgoing electric currents. It also explains the presence of excess radiant energy, according to the formula $E = mc^2$, which may be written $m = E/c^2$ and now may be modified to $m = E/c$, as explained later in Part Three of this book. In these equations m equals the mass added to the earth's materials, E is incoming radiant energy, and c is the speed of light. This excess radiant energy is continuously building up the top stratum of the surface of the earth with new materials, as explained in Part Three, "Origin of the Earth's Materials."

One of the fundamental tenets of the Drag of Gravity theory holds that the surface of the earth contains a quantity of electricity known as static electricity. Yet, since any conducting body on the earth's surface, charged with like electricity and separated from contact with the earth, would tend to be repelled at once, it becomes evident that the Drag of Gravity is a tremendously greater force than the force of electrical repulsion. The forces of the incoming radiations are so powerful that the forces of electrical repulsion seem to become infinitesimal. The Drag of Gravity thus appears to produce a binding force on the atoms of the earth's materials.

As an analogy, let us consider the structure of the atom. The nuclei of the atoms are understood to be composed of protons, all of which are electrically positive and which ought, therefore, to forcibly repel one another. But a binding force—greater than the force of electrical repulsion— welds the protons and the neutrons of the atoms into a single stable kernel.

Ever since the Brownian movement of molecules was discovered it has been known that molecules are in constant motion in liquids and gases and that they bump against one another. The positive protons and negative electrons in atoms and molecules are electrical entities that revolve and spin, and they thus create electrical currents with coexisting magnetic fields; fields of vari-

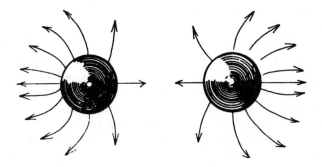

FIG. 13. Static attraction and repulsion of electrically charged conducting bodies. Two insulated balls are both charged with the same kind of electricity—either positive or negative electricity. The tensions along the lines of force now tend to push the electrifications on the surfaces away from each other, decreasing the electrical charges on the near sides and increasing the electrical charges on the far sides. By opposing each other on the near sides a super-saturated condition of the intervening medium is created which pushes the balls apart.

Static Electricity

ous neighboring particles merge, the consequence being that all atoms and molecules are characterized by a pattern of interacting magnetic forces. The primary forces producing these motions and effects are the incoming radiant energy rays which strike them. The outmoded Attraction of Gravity theory cannot account for these motions.

Physical experiments show that insulated spheres, being electrical conductors, have the ability to retain charges of electricity —subject to slow leakage into the surrounding atmospheric medium. The charges reside solely in the outer surfaces of the spheres, whether the spheres be solids or hollow shells. The electrical charges may be imparted to the spheres by contact with another electrified body, or by "influence" of a relatively close electrically charged body which does not touch it. An insulated conducting body has not only the property of holding a charge of electricity, but when charged it fairly bristles with electricity. We cannot see this "bristling," but we can measure the relative amounts of the electric charge at different distances.

We observe that two bodies charged with like electricity *repel* each other. The fact is, however, that the force of repulsion is in the intervening medium and is not an inherent property of the bodies themselves. It is the stresses in the intervening medium which cause electrically charged spheres to move relative distances from each other. A supersaturated intervening medium pushes the charged spheres apart when both spheres are charged with like electricity.

We also observe that two bodies charged with unlike electricity attract each other. We say they are attracting each other because we see them come together.

The fact is that when opposite types of electricity in the two spheres neutralize each other, an electrical vacuum is created between them. The electrical bristles of one become paired off and neutralized by the contrarily charged electrical bristles of the other. In this electrical vacuum, the pressure of the ether surrounding the two spheres pushes them together. Thus, a neutralized intervening medium allows two spheres to come together.

We have been observing that an imponderable force causes

Fig. 14 A. A represents an insulated ball which has been charged with electricity—either positive or negative. The electrical lines of force are simply radial. The electrical charge is uniformly distributed over the surface.

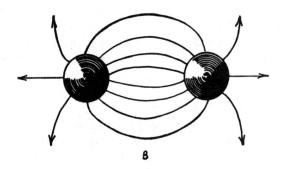

Fig. 14 B. B represents two similarly electrified insulated balls which are oppositely charged, and brought near to each other. One has a positive and the other a negative electric charge. There is a tension along the lines of force so that they tend to draw the electrifications on the surfaces of the balls toward one another, increasing the electric charges on the near sides and lessening the electric charges on the far sides. By neutralizing each other on the near sides a vacuum-like tension in the intervening medium is created, drawing the two balls together. There is also a lateral pressure in the medium tending to keep the electrical lines apart from one another. These lateral pressures also cause an unequal distribution of the lines over the surfaces. They are densest on the parts nearest each other.

Static Electricity

the observable motions of ponderable substances. We also observe that when the two spheres touch together—and thus equalize their electrical charges—there is no longer any electrical stress in the intervening ether.

It is as natural to say that two spheres attract or repel each other as it is to say that the sun rises in the east and circles around the earth. The facts are that the earth itself rotates and the sun is relatively at rest. Similarly, the two spheres which seem to be attracting or repelling each other are actually being pushed around by the electrical ether which is invisible.

It is, therefore, not the spheres that attract or repel each

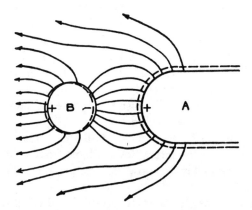

Fig. 15. When a nonelectrified metal ball B is brought under the influence of a positively electrified body A the action is one in which the intervening medium takes an essential part.

Some of the electrical lines of the field that surrounds A pass through B, entering it at the side nearer A and leaving on the far side. (The distributions of the electrical charges on A and B are shown by dotted lines.)

If the ball B has no charge of its own, as many electrical lines will enter on one side as leave on the other. The induced negative charge on one side and the induced positive charge on the other will be exactly equal in amount.

other, but the electrical charges that they carry which determine their positions in relation to each other.

Consider, for example, a large, circular room or laboratory with a high ceiling, in which a hundred pith balls, varying in size from ½ inch to 3 inches in diameter, are suspended from the ceiling by silk threads in such a way that they hang at a height just above a man's head; they are all in the same horizontal plane, and are irregularly spaced in relation to one another.

Then, assume that the air in the room is very dry and motionless and that each pith ball is electrically charged, from contact with a cat's fur which has been rubbed on glass or any other material that will impart an electrical charge.

Due to the electrical stresses in the intervening ether, the pith balls will then repel each other, like an expanding universe.

Now, assume that a number of gamma sources are placed, at random, around the walls of the laboratory so that they project gamma radiations in the plane in which the pith balls are freely suspended on their insulated threads. The force of these rays striking the pith balls will cause the spheres to move about, but the pith balls will never collide.

This is, of course, theoretical. Radiating pressures have been demonstrated, but the intensities available from most present-day sources are probably too low for such a quantitative demonstration.

When we relate to the solar system the physical facts that we have observed in regard to pith balls, it becomes evident that the electrical stresses in the intervening ether cause spheres, sun, earth, moon, planets, stars, all of which are free to move, to arrange themselves in accordance with these stresses. The stresses cause the motion. The spheres are held apart at "arm's length," the stresses in the ether being the arms.

All the planets of our solar system revolve around the sun in a west-to-east direction. Our moon and the two moons of Mars revolve around their planets from west to east.

Jupiter has twelve moons. The eight inside moons travel from west to east, but the four outer moons revolve "the wrong way," from east to west. Retrograde or east-to-west motions are also

Static Electricity

characteristic of the outermost satellite of Saturn's nine moons, of four moons of Uranus, and of one moon of Neptune.

When these motions of the satellites—some in one direction and others in the opposite—are assumed to be due to electrical repulsion from all other heavenly bodies, we arrive at the simplest and most readily understandable answer to what is otherwise a paradox; except for the existence of the laws of dynamic electrical repulsion there is no reason why the satellites of the planets should travel exactly as they are observed to be traveling.

The sun keeps its planets and the planets keep their satellites at arm's length through static electrical repulsion. Dynamic electrical repulsion causes the observed revolutionary movements; they can be most readily explained as being caused by the impacts of the radiant energy radiations which fill all space.

It appears that it is these dynamic radiations which produce the motivating gravitational fields of planets as well as satellites —just as celestial electrical radiant energy rays produce the Drag of Gravity at the surface of the earth, causing the earth's diurnal rotation and the weights of its materials.

The elliptical orbits of planets and satellites are also readily accounted for when we keep in mind the fact that action and reaction are always equal and opposite in direction. For example: the static electrical repulsive force from the sun to the earth and also, the one from the earth to the moon are constants at any fixed distance, but vary with the distance separating the bodies—inversely as the square of the distance.

At any moment and at any point along the orbits of earth or moon, the opposing forces of static electrical repulsion, on the rear sides, must be exactly equal to the sun's and to the earth's static electrical repulsive effects in order to produce the resulting, and existing, positional equilibrium of the earth and of the moon in the solar system.

The opposing static electrical repulsions obviously must vary with the changes in the earth's and the moon's positions in the universe, as they travel through space, because these positions, at any moment, determine what repulsive forces are pushing on the opposite or rear sides. Therefore, as the opposing static repulsions vary, the observed result is that the paths of travel of earth and moon, in their orbits, become distorted circles—resulting in elliptical forms.

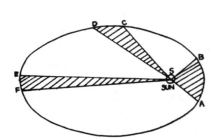

In the Copernicus diagram, showing changes of speed of the earth in its orbital travel, the ellipticity of the earth's orbit around the sun is purposely shown on a greatly exaggerated scale. The variations of the orbit from the average distance from sun to earth is only 1.67 per cent.

FIG. 16. Changes in the earth's orbital speed. The shaded areas shown, SAB, SCD and SEF, have equal areas.

The earth's speed in its orbit around the sun varies. The time from A to B, from C to D and from E to F is the same for each distance, illustrating the greater speed from A to B than from C to D or E to F.

The speed of the earth in its orbit around the sun changes from fast to slow as the distances between earth and sun increase, and from slow to fast as the distances decrease during the year.

The operational result of these positional cosmic forces is that the orbit of the earth deviates from an average or perfect circle by an ellipticity of 1.67 per cent, and the orbit of the moon by 5.49 per cent.

It is here postulated that the earth's rotation is caused by dynamic energy radiations and that its revolutions around the sun are caused by static electrical forces. The revolutions of the earth around the sun and of the moon around the earth are in approximately parallel planes. The plane of the moon's orbit around the earth varies approximately 5° from that of the earth's orbit around the sun—referred to as the ecliptic.

Static Electricity 201

The rotation of the earth occurs on its Axis of Figure, which is slanted to the ecliptic by 23° 27'. It becomes evident that different imponderable forces are responsible for the different directions of motion; if it were the same force in each case, the revolutions and the rotation would be in more nearly parallel planes.

The speed of the earth in its orbit around the sun varies. The earth moves faster when nearer to the sun, and proportionately slower when further away. The diurnal speed of rotation of the earth—compared to a perfect clock—varies by 8 or 10 seconds, being fast or slow at rather irregular intervals of a few decades. It therefore becomes obvious that the earth is not spinning from the momentum of a force imparted to it once for all at a time

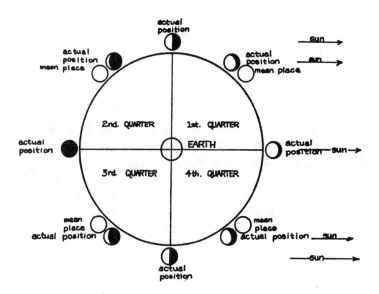

Fig. 17. Changes of speed of the moon in a twenty-eight-day period. The moon, in its journey around the earth, is ahead of its mean position in first and third quarters and is behind its mean place in second and fourth quarters. This change in speed is called the moon's variation.

called "Creation", but that it is spinning due to constantly applied forces of radiant energy, its total intensity being subject to slight variations.

The moon moves fastest in its first and third quarters, when it is moving away from the earth along its own orbit. It moves most slowly in its second and fourth quarters, when, in moving along its orbit, it is again approaching nearer to the earth.

This confirms the theory that static electrical repulsion fixes the distance between earth and moon; the speed of the moon's motion is greater when it doesn't go counter to that repulsion, and it is less when the motion and the forces of repulsion go in opposite directions.

The relatively small variation in speed indicates that the disturbance of the intervening ether between earth and moon, by the electrical charges carried by these bodies, is very minor compared to the tension in the ether caused by all the electrically charged spheres of the universe. It is that greater tension which mainly determines the relative positions of earth and moon.

The evidence shows that it is repulsion and not attraction that causes the changes of speed in the moon's motion.

The so-called evection has a period of 1⅜ year and displaces the moon by 1¼ degree of arc forward or backward. The so-called variation has a period of a month and displaces the moon by 40 minutes of arc in such a way that the moon is ahead of its mean place by that amount between new moon and first quarter, and between full moon and third quarter—and behind at other times.

Furthermore, as the moon moves through space the directional path of its motion is continually changing; its motion therefore cannot be due to momentum, which moves a body only in a straight line. Two reasons for its zigzag, serpentine, or wavy motion, are (1) that the moon's orbit around the earth is inclined about 5° to the orbit of the earth around the sun, and (2) that the Axis of Spin of the earth is inclined 23° 27′ to the plane of the orbit of the earth.

Therefore, as the moon journeys through the heavens, while constantly revolving about the rotating and revolving earth, it rises and falls, like the horses on a merry-go-round, rotating on

Static Electricity

a 23° 27′ slant, with a 5% variation, while at the same time it moves forward in space with the whole solar system; this results in a curvilinear motion, not a straight line, and eliminates the theory of motions by momentum.

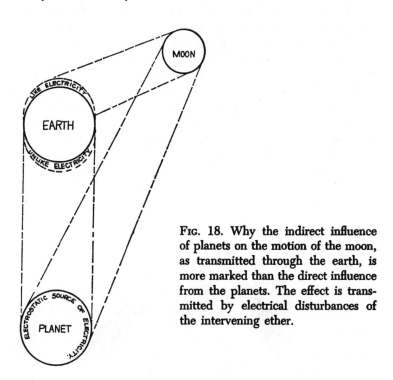

Fig. 18. Why the indirect influence of planets on the motion of the moon, as transmitted through the earth, is more marked than the direct influence from the planets. The effect is transmitted by electrical disturbances of the intervening ether.

It has been observed and recorded that the direct influence of planets or the motion of the moon is much smaller than their indirect influence as transmitted through the earth.

Obviously this cannot be caused by the "attraction of masses," but would naturally follow the laws of electrical influence and confirms the assumption that all heavenly bodies are charged with "like" electricity. Like repels like.

The earth, being a very much larger body than the moon, receives more electrical effect than the moon from the approach or retreat of a planet, and the electrical stresses and disturbances caused in the ether by the changes in volume of induced "like" electricity are a greater influence on the position of the moon than is the direct electrical influence of the planet.

Comets on their first appearances always have their tails pointing away from the sun. When a comet passes from one side of the sun to the other side, its tail changes its observable position, and, whereas the tail followed the comet when approaching the sun, the comet follows its tail when receding from the sun.

If the head of a comet is solid matter, its place with its tail in celestial space is fixed by static repulsion, like other heavenly bodies. The tail—being of less solidity, possibly gaseous—is deflected away from the sun by the bombardment of the sun's rays, like a feathered shuttlecock in the wind.

The weight of evidence based on the foregoing physical facts

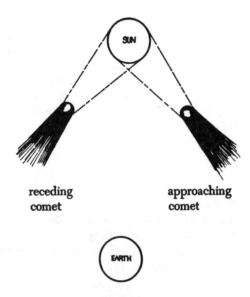

Fig. 19. A comet's tail follows as it approaches the sun. The comet follows the tail as it recedes from the sun.

Static Electricity

—all refuting the theory of universal mutual attraction of masses— is sufficient to prove that the theory is unscientific, being in variance with observable facts, and leads to the only possible conclusion: *Universal Mutual Attraction of Masses Does Not Exist.*

Tides

STATIC REPULSION causes the tides of the oceans to be low under and opposite the moon. When the tide tables given in the publications of the United States Coast and Geodetic Survey are compared with moon tables for same day and hour, they show low water under and opposite the moon at various ports open to the ocean and not in estuaries. Science writers generally adhere to the postulate, but now believed to be erroneous, that every particle of matter in the universe attracts every other particle of matter to itself. Obviously, such mutual attraction of celestial masses for each other would cause the universe to collapse. Astronomers studying the motions of stars and galaxies tell us that the universe is expanding and is not contracting. Obviously, an

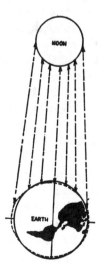

FIG. 20. Static electrical repulsion between moon and earth causes the liquid portion of the earth to be depressed when under and opposite to the moon, resulting in the diurnal tides.

expanding universe and the mutual attraction of masses within it cannot co-exist. An expanding universe and low water under and opposite the moon both indicate that electro-static repulsion is the motivating force.

Ocean water levels vary with the electro-static repulsions of moon and sun (we here leave out wind and storm pressures). The moon's pressure, which is about 2½ times greater than that of the sun, may be aptly called its aura. Both cover about half of the surface of the earth at any one time. Since water is practically incompressible, the electro-static pressures are immediately and uniformly distributed. The greatest depression of the surface waters of the oceans occurs directly under and opposite the moon, and the least effects of moon pressures are found to occur at the edges of its aura, causing the phenomena of low water under the moon and high water about six hours later. High tides rise higher when barometric pressures are low, a fact which confirms pressures as controlling tide ranges.

At the latitude of New York City the diurnal tidal trough, under the moon, moves with the speed of the rotation of the earth, at approximately 600 miles per hour. Traveling at that speed the tidal trough arrives at the same place the next day. The moon remains overhead, but travels faster than the surface of the earth, so it takes approximately 52 minutes each day for the earth to catch up to its former position under the moon, varying from 13 to 80 minutes. About six hours later the tidal protuberance, following the trough, arrives at the same location.

Sandy Hook is on the west shore of the Atlantic Ocean, and its tidal changes are not affected by their having to pass through estuaries. Tide charts and moon charts show low water under and opposite the moon at Sandy Hook. This is one example, easy to check for accuracy, that confirms the fact that static repulsion, and not attraction of masses, is the cause of the fall and rise of the ocean tides.

At Sandy Hook, N.J., where the Atlantic Ocean joins New York Bay, a secondary or estuary tide takes over. At the times of flood tides it reaches the Battery at New York in 35 minutes at a speed of 30 miles per hour (for tide level, not current flow). The

Static Electricity

tide wave then continues up the Hudson River to Poughkeepsie, at a speed of about 16 miles per hour (See "Time of Tides on Atlantic Coast," *World Almanac*). The rates of current flow of tidal waters in the vicinity of New York City varies from zero to about four miles per hour.

All celestial bodies are here postulated to be charged with "like" electricity. Because like repels like, the universe is expanding, and is not contracting from the assumed attraction of masses. Electro-static repulsion is what holds the earth securely in its place in the universe. A rubber ball, rolled under the flat palm of the hand, offers an analogy for action and reaction. The moon and sun furnish action on one side of the globe and the stars of the celestial heavens, above the opposite hemisphere, furnish the reaction. The repulsions of moon and sun are a small part of the total forces of celestial repulsion, as otherwise the tide ranges would be greater. Electro-static repulsion has been demonstrated as a true law of nature, while attraction of masses has failed to meet the acid test of duplicate measurements.

Earth tides in water wells have been extensively researched and show low water in wells at the time of the moon's transit, indicating that the whole earth is affected by the moon's aura.

From a preliminary study of the time of tides at ports on the Atlantic and Pacific Oceans, compared with the moon's transit, the theory of static repulsion on the ocean waters is confirmed and the theory of "moon pull" is confounded. Research indicates that few tidal anomalies should remain when careful studies are made of the speeds of estuary tidal waves, and the effects on the tidal waves of depths, contours, and coastal shelves of the estuaries.

The closer the moon and the sun get to the earth, the greater the tidal disturbances. The lowest and highest monthly tides occur when perigee (the time when the moon is nearest to the earth) and new and full moon (the time when the moon and the sun are on the same or opposite sides of the earth) occur together, and become still lower and higher when the earth is nearest the sun (perihelion), as it is in January.

It is evident that the ocean waters are not being pulled up

as the earth rotates under the moon; if they were being pulled up into a protuberance, then surface currents would disclose the flow of water. On the contrary, when the ocean waters are compressed, by repulsion from the moon, then subsurface currents, difficult to detect, are created and promote west-to-east surface currents as the tide trough moves westward. H. U. Sverdrup, in his classic work *The Oceans* (page 551), states that "The obvious criticism that can be directed at this theory [of lunar attraction] is that a movement of a flood protuberance over the surface of the earth cannot take place unless water masses actually change positions; but consideration of the movement of the water has been completely disregarded" [by a list of authors in a voluminous bibliography].

Physical Evidence Supporting Gravitational Repulsion

THE FORCE OF gravity varies a small fraction of one per cent at different locations on the earth's surface. We notice that some differences in the force of gravity occur relatively close together. At some locations near great mountains the force of gravity is less than at some locations on the plains. This is a direct blow to the theory of universal mutual attraction of masses.

The widely believed theory that the force of gravity is greater near large masses, such as mountains, is confounded by these recorded observations of actual measurements of gravity.

In the reports of measurements of the force of gravity in India, by the British Survey, the theory of "the hidden range" has been introduced to account for some of the variations from that accepted theory. No such invention is required for the Drag of Gravity theory to explain the variations in observed measurements of gravity.

It should therefore be clear that the theory of universal mutual attraction of masses does not fit the facts disclosed by the measurements of the force of gravity at different locations on the earth's surface—while, on the other hand, the Drag of Gravity

Static Electricity

theory does fit the facts disclosed by these same measurements.

The gravitational effect of the sun or of the moon at any point on the earth is at maximum when the sun or moon is nearest to it, and is also at maximum at the same time on the opposite side of the earth, at the point farthest away.

Fluctuations in the force of gravity—caused by the relative positions of the sun and the moon—have been measured by gravity meter and found to vary with great regularity during each day. When plotted as curves, the hourly readings are nearly identical for numerous days in succession.

When the curves for daily variations in gravity are assembled for longer periods of time, the over-all curves are found to rise and fall. These have been called gravity drift curves. Both the daily curves and the drift curves follow the variations in the distances from the moon and the sun to the points of observation on the earth.

Earth tides in water wells and mine shafts have been analyzed and show low level at the time of moon's transit (Special Publication #223, U.S. Coast and Geodetic Survey, *Report of Earth Tides 1936-38* by Walter D. Lambert).

The literature on gravitation reveals that many erudite scholars prefer to hold fast to what they were told in their youth, the belief that gravitation is an attraction between material bodies. The advancement of science is impeded when these scholars refuse to acknowledge the truth of gravitational repulsion by dynamic and static electrical forces.

VI

Research Projects

WHEN A NEW theory is asseverated, tests of its fundamental accuracy need to be obtained even though the theory seems to be entirely reasonable. To this end certain challenges are offered, citing physical effects which scientists could not at the moment predict with confidence, but which will naturally follow during such tests, provided the theory is correct. Such tests, if made and reported, will confirm or refute the theory of the Drag of Gravity.

We here list a number of suggestions for measurements which can be made with a fair degree of accuracy, and which relate to physical effects which could not be predicted with confidence under the prevailing theory of universal mutual attraction of masses, but are expected to confirm the Drag of Gravity theory.

Research Project No. 1

AFTER REACHING a maximum, pressures do not increase arithmetically with greater depths in the ocean.

According to our accepted rules of hydraulics, a column of sea water one foot high exerts a pressure of .44 pounds per square inch, and the pressures increase uniformly at .44 pounds for each foot until the bottoms of the oceans are reached.

At three miles down, the pressure (according to the formula) is over 7,000 pounds per square inch—or over 1,000,000 pounds per square foot—which is a value far above the crushing strength of such building material as brick. ("The crushing strength in lbs. per sq. in. of . . . common red bricks . . . ranged . . . 3010

Research Projects

... 4080 ... 4960 ... 6361 ... Ordinary granite ranges from 20,000 to 30,000 lbs. compression strength per sq. in." Kent, *Mechanical Engineer's Pocket Book*).

Yet from a depth of three miles—where the above great pressures are commonly assumed to exist—"bottomfish" and worms have been brought alive to the surface. (See reports from S.S. *Challenger* in *Depths of the Ocean*, by Murray and Jhort, p. 418.)

This supports the theory that gravitational pressures are surface pressures, for animal life could not exist under the commonly conceived bottom pressures and such animals also remain alive when brought to the surface, where of course such pressures are absent.

At depths of five miles—in the Pacific Ocean (where pressures, according to the formula, are 11,616 lbs. per sq. in., and should have changed the sea bottom to solid material)—a dredge produced slime, muck, meteor dust, etc. Depths of seven miles have been sounded (having 16,262 lbs. per sq. in. pressure according to the theory).

Proof or refutation of these great but theoretical bottom pressures is called for. The Drag of Gravity theory calls for their refutation.

Project No. 2

A TEST WEIGHT, on an accurate spring balance, will show a diminution of weight in a deep mine—due to the shielding effect against celestial radiations of the earth materials above it. The percentage decrease in weight will be much greater than the percentage that the distance to the surface bears to the radius of the earth.

Project No. 3

A TEST WEIGHT, on a spring balance, in an anchored balloon or in a stationary dirigible airship, will show a diminution in weight not in proportion to the square of the distance from the earth, but in proportion to the loss of shielding effect against celestial radiations due to its distance from the earth.

Project No. 4

IN ASTRONOMICAL calculations much of the mathematics of repulsion and attraction may prove to be interchangeable, for Newton used the word "attraction" in his *Principia* to include either or both. Therefore, when a mathematical approach to calculations of variations in the speed of the moon, for example, as it approaches and recedes from the sun, is based on the theory of repulsion, it will be expected to encounter fewer difficulties than have mathematical scientists relying on the theory of attraction.

Project No. 5

A SHIP SAILING in the ocean toward the moon is sliding down an inclining plane, and should go faster; similarly, when sailing away from the moon, it goes uphill and should go at a slower rate. If those interested in predicting ships' positions by dead reckoning, will take this fact into consideration, including the west-to-east gravitational push among others, they may be able to arrive at more accurate predictions. If so, their voluntary reports from time to time should prove helpful to science.

Project No. 6

MEASUREMENTS OF the shielding effect of large masses, such as mountains, against celestial radiations in one direction should show plumb bob deflections, and may be discovered by measurements made by either mechanical devices or electrical instruments.

Project No. 7

ALSO TO BE desired are laboratory proofs of celestial rays causing the weights of materials. One experiment would be to endeavor to shield a test weight on a spring balance from all celestial rays

Research Projects

coming from a given direction; or, to produce a celestial ray vacuum while maintaining normal and undisturbed barometric pressures.

Strong electric currents, like artificial lightning, should deflect incoming celestial rays sufficiently to show instantaneous changes in weights of test pieces. Extremely thick lead roofings may show a similar shielding effect.

Project No. 8

THE VARIATIONS IN the force of gravity at different locations on the earth's surface—particularly where there are sharp differences relatively short distances apart—should be studied and investigated for the purpose of establishing proofs of the existence of radiating materials beneath the surface. *The Drag of Gravity theory recognizes a radiating earth!*

Project No. 9

SPRING SCALES HAVE been referred to as "notoriously inaccurate" because they measure true weight and not relative weight. True weight varies, so that an accurate spring scale will show different weights at different times for the same materials.

Changes in the weights of materials as the relative positions of earth, moon, and sun change or as atmospheric voltages change, have not yet been recorded nor reported; however, repulsion due to direct sun rays is a matter of record, and daily variations in the electrical voltage of the atmosphere is also a matter of record.

Variations in the weight of a test piece of material should be discoverable, when measured on a spring scale over any twenty-four-hour period of time and during periods of electrical disturbances, when cosmic radiations make the flow of earth currents sufficiently large to interrupt telegraph and telephone services.

PART THREE

ORIGIN OF THE EARTH'S MATERIALS

> We must learn that any person, who will not accept what he knows to be truth, for the very love of truth alone, is very definitely undermining his mental integrity.
>
> LUTHER BURBANK

I

Origin of the Earth's Materials

As the autumn leaves fell at the end of a five-week rainless period, the edges of the pavements of the streets of suburban New York City held layers of thoroughly dried-out, brittle leaves that crumbled to a fine yellow powder of desiccated vegetation; a light rain then washed the finely granulated materials into the gutters, where they changed into mud. I took pains to examine it and found that it was no longer a yellow powder, but had become a dark, slimy mud. When the mud dried out, it became dirt. Dirt becomes hardpan, and hardpan becomes rock.

Layers of black, wet mud, exposed at low tides along the shores of Long Island Sound, consist mainly of organic matter that once floated in the waters and then sank to the bottom. On the other side of New York City, the large area of the Hackensack Meadows is made up of dark brown, compacted mud, consisting mainly of disintegrated vegetable matter that during the past 7,000 years first floated and then sank in the waters, changing a once shallow esturary tidewater lake into meadow lands.

Botanists inform us that all vegetation consists of about one per cent mineral matter which is left behind, like ashes from a fire, when the vegetation dies; its other ingredients change into water vapor and carbon dioxide gas and escape into the atmosphere. This gives us an elementary picture of how minerals develop on the earth and why the earth has been constantly growing larger during approximately four-and-a-half billion years.

Physicists inform us that matter consists of molecules that are composed of atoms in constant vibration, each atom being made up of a nucleus charged with positive electricity and surrounded by negatively charged electrons in circular motion. A piece of steel is composed of submicroscopic elements in constant motion and these expand when heated because the atoms then move faster and take up more room.

Spectroscope operators show us that each of the 92 basic chemical elements, every material of the earth, can be identified by the wave lengths of its light rays when made incandescent or is burned, at which moment the materials change into gases and light rays. The spectroscope intercepts the light rays as they return to space from which they originally came, and identifies the material by the wave lengths of its light rays.

The chemical elements of the materials in rocks which were once leaves and vegetation, can be identified by the light rays that created them; we thus assume that they all came from celestial space in the form of light rays, and that the atoms of the light rays became frozen or occluded and changed into the atoms in the materials of the earth. The process is called photosynthesis—the synthesis of chemical compounds effected with the aid of radiant energy, especially light, and specifically, the formation of carbohydrates in the chlorophyl-containing tissues of plants exposed to light.

Coal is a mineral rock that was once vegetable matter and is found sporadically in all parts of the world, at different depths in successively created strata of the earth. The minerals silica and calcium are now found associated with former vegetation in the form of petrified trees and granitic rocks. Silica sands and calcium limestones are created in the oceans. The association of these minerals can be explained as being one result of the recurrent great cataclysms of the earth; these occurred when polar areas, overburdened by ice, rolled to the tropics and tropical areas were moved to the poles, and, they marked the end of that particular epoch of time.

Research appears to confirm that the sun, stars, planets and meteorites, comprising the entire universe, are composed of the same ingredients and that these ingredients sometimes appear

Origin of the Earth's Materials

as energy and sometimes as matter. Energy and matter have been proven to be mutually convertible. We have abundant evidence that energy from celestial space changes into matter. The coal beds which are now underground organic rock layers are one example. What is now rock was once energy which was transmuted into vegetation and then metamorphosed from vegetation to rock. Petrified wood, leaves, figs, and animal bones have gone through a similar process, having first been created on the surface of the earth, and then been metamorphosed to mineral rock while buried underground.

Television, radio, and electronics give us clues to certain processes of creation. In television we have an *artificial* transmitter of electrical oscillations and a receiver for utilizing them. Certain forces of nature are harnessed and utilized, by the intelligence of man, to produce pictures on a screen at great distances from the sources of their creation. The oscillations are broadcast, like the radiations from the sun.

A blade of grass grows in a field because an intelligence, far superior to man's intelligence, has caused the creation of radiant energy which creates blades of grass, at great distances from the energy sources.

Neither the light rays nor the radio waves are visible; but they are just as real as the blades of grass and the television pictures which we see and which are the result of the invisible energy radiations. These energy radiations created them and caused them to become visible to our eyes, whenever captured by the receptive medium.

Consider now the enormous volume of energy being radiated by the uncountable number of stars in celestial space, all of which radiate energy just as our sun throws off energy radiations. This energy arrives at the surface of the earth in a continuous shower. Earth-bound electrical or ethereal images convert certain of these invisible forces into materials, by photosynthesis, under suitable conditions of temperature, humidity, barometric pressure, and a suitable electrical condition of the atmosphere.

Nature creates blades of grass in a field. Man creates television pictures on a screen. Both the blades of grass and the television pictures indicate to us the existence of creative intelligence. The original forces of nature were in existence a very long time before men learned to utilize them in order to produce television pictures on a screen. The creation of the television pictures provides us with an analogy of the invisible forces which create the blades of grass; but the methods used are quite different.

The 92 basic chemical elements of the earth (*now sometimes listed as 102*) are identified by the lengths of the light rays into which they metamorphose on being made incandescent. These chemical elements arrive on the earth as energy rays and return again to celestial space as energy rays.

Based on the latest findings of physical and cosmic science it seems reasonable to postulate that everything in and on the earth, excepting its core, has been created within the past 4½ billion years—including land, oceans, and atmosphere. These three entities are each composed of some or all of the 92 basic chemical elements, whose essences all become individual light and heat radiations and return to celestial space at temperatures above incandescence in the case of minerals and metals, and at hydrogen bomb temperatures in the case of certain gases.

Consider the topsoil of the United States some 50,000 years from now, when it may be metamorphosed into solid rock and buried, say, 100 to 400 feet below the surface of the earth. Thick rock would mark the bottoms of present valleys and river deltas. Thin rock would mark the areas of some of our highlands, while the bottoms of swamp-lands, such as the Everglades and Dismal Swamps, would have become peat or coal.

Elements of Continuous Creation

WE WILL REVIEW briefly what is known and suggested concerning the growth of vegetation and layers of inorganic materials.

The germs of vegetable growth are the male and female life forces which develop in stamen and pistil respectively, and when joined together in an environment suited to the species, create

Origin of the Earth's Materials

zygotes. These, in turn, create cells or miniscule bodies, and plants grow by cell division at upper terminal of stem and lower ends of roots. The individual cell, once formed, does not move upward. The cells created by the zygotes consist of atoms, which makes it reasonable to suggest that the zygotes create the molecules and that they have the function of attracting and binding electrical ions of the air with photons from celestial space. The photons are corpuscles of light, mainly from the sun. The ions are absorbed from currents that flow from stem to root and are caused by the difference in electrical potential between air and ground. The omnipresent ions create the voltage gradient of the ambient air, and moist earth creates the necessary conducting electrical contact between root and soil.

The zygote cells have the unique property of subdividing; then, as each subdivision grows to the size of the original cell it again subdivides, and this process continues indefinitely—but only within limiting temperatures. The zygotes of each succeeding cell are the life forces of vegetation. The ions, which are captured or occluded when flowing through the live parts of the plant, are postulated to become the protons and neutrons of atoms, whose whirling electrons are the captured corpuscles of light. These known ingredients of air and sunlight account for the continuous creation of the atoms by metamorphosis. The end use of most of the tremendous quantity of energy from celestial space, which is being rained continuously upon the surface of the earth, is thus accounted for, and the equivalence of energy and matter is confirmed.

The primary forces of nature are identified as radiant energy acting on electrical entities which gather and hold matter. The prime creations become identified as meristomatic cells, auxines, enzymes, and growth hormones, as observed in plants. Each develops along an astral or ghostlike pattern, which manifests its presence visibly when clothed with the materials which it holds together, as presently described for the snowflake. Every atom of each material of the earth is an electrical entity of earthbound energy, which has been occluded, captured, or frozen in its transient material form.

The metamorphosis from radiant energy to matter is being observed constantly where microscopes reveal the growths of the cell structures of plant and animal life, and electrical instruments reveal the concurrent voltages of the cells; the source of their growths is thus identified as being an unseen all-pervading energy supply.

Chlorophyl is the medium in which electromagnetic forces of photosynthesis convert the photons and quanta of light rays, by a form of metamorphosis, into atoms of the earth. Atoms unite to form molecules, which are grouped along the astral patterns or ghosts of the creating zygotes, and nowhere else. By analogy the astral pattern is somewhat like an invisible but necessary molecular hatrack, whose form varies with the species. The building materials include the invisible life forces, light corpuscles and ions, and the visible materials which are carbohydrates plus some inorganic materials. All vegetation has an average content of approximately $C_6H_{12}O_6$ with about one percent of minerals.

When vegetation dies, the carbohydrates are consumed by slow combustion (rotting) and are metamorphosed into carbon dioxide gas and water vapor. The CO_2 is re-used in creating new vegetation. The H_2O condenses as rain and adds to the volume of the oceans, as explained presently. The minerals are left behind as a residue or ash and are found throughout the successive earth strata in which they were created.

Mineral matter, found in all vegetation, consists of calcium, iron, copper, and other inorganic materials. Calcium is being created continuously in vegetation and appears in the limestone strata of the earth. Sandstone is created by sand crystals of silica, which is a direct fixation of photosynthesis, occurring only in shallow seas and never at depths below which the sun's rays penetrate. The formations of crystals in the air, snowflakes, is an analogy. Each of these citations will be presently explained in further detail. Granitic rocks and petrified trees are developed underground by silica and calcium carbonate, which crystallize with the buried vegetation to form petrified forest floors and trees, after having been picked up and carried in tenuous solutions of rain water percolating through the overlaid strata. A part

Origin of the Earth's Materials

of the vegetation, buried by the cataclysm that ended one epoch, is returned to the surface, in the form of rock, by one of the succeeding convulsions of the earth.

The snowflake is an example of electrical forces promoting the binding together of materials. The ions of the air are recognized as the cause of the voltage gradient of the ambient atmosphere, as outlined later in the chapter "Volcanoes and Hot Springs." They hold together the water in raindrops. The condensation of water vapor in dust-free air is demonstrated by the Wilson cloud chamber.

All snowflakes which fall to the earth are electrically alive until the voltages of the ions are neutralized by becoming grounded on the earth's surface. Microscopic water particles have become electrically charged by the ions of the air, causing them to have north and south poles. The south and north poles attract each other. The south pole of a microscopic moisture particle then freezes fast to the north pole of an adjacent particle, and they become visible when enough of them have collected on the ion. The beautiful astral pattern, or ghost, of the ion then becomes visible because the moisture particles are ionized and gathered only along the straight lines of this astral pattern, now clothing itself with frozen moisture and becoming a snowflake. The snowflake is a mineral crystal which, like all other elements of the earth, exists only within a given temperature range. Water is one of the minerals of the earth.

The identity of the ions of the air with snowflakes has been proven true by reports of St. Elmo's fire leaving the wing tips of planes as well as propeller blades when airplanes fly through snowstorms. The ions arrive on the planes as snowflakes and their flow-off as St. Elmo's fire confirms their being electrically charged. Due to the excess of static, radios of airplanes become useless during snowstorms. The static is created by the Hertzian waves being generated as the ions of the snowflakes become grounded on the bodies of the airplanes.

Atoms, when grouped as molecules and occluded or frozen as matter, make up the chemical elements of the earth's materials. They are identified by their wave lengths on spectrograph charts, as previously explained. All of the earth's materials are transient, occluded, earth-bound forms of the stellar radiations which fill all space. Energy radiations, atoms and ions, are the same thing in different manifestations. The growth of the earth and its stratified construction are accounted for by this new cosmic, geophysical theory of continuous creation, caused by electrodynamic interactions of radiant energy from celestial space and earth-bound energy forms.

This explanation identifies the earth's materials with the rest of the universe, all being the same essence in different natural manifestations. It explains how each of the successive strata of the earth represents, generally, the accumulated growths of vegetable, mineral, and animal matter, created during each one of the epochs of time punctuated by the successive cataclysms of the earth.

The growth of upland materials varies with their locations—being greater in temperate and tropical zones than in frigid zones; while few data have as yet been gathered about the rates of growth of slimes, mucks, corals, limestones, and sands of the sea bottoms.

The earth is growing larger in diameter at the rate of approximately four feet in each thousand years. The average build-up of topsoil for any one place is about one foot in each five hundred years. This approximation is based on the data of the drilling at Spur Ranch—as already given in detail. (See page 73). The average stratum was found to be about thirteen feet in depth.

It is assumed that each stratum represented one epoch of time, each of which averaged six thousand years for the upper strata; this figure is subject to correction when better data on the duration of epochs become available.

At Ur of the Chaldeans and at Cnossus, Crete, where excavations have been made, the build-up is found to be six or eight feet for each thousand years. Those underground remains of cities clearly show how the development of topsoil materials have buried the evidences of the various civilizations.

Entirely new concepts of the formation of some of the earth's

Origin of the Earth's Materials

materials will follow, as naturally as day follows night, when we accept as an axiom in geophysics the fact that the earth rotates on different Axes of Figure at recurrent intervals of 6,000 to 7,000 years. The metamorphosis of the rocks of the earth will be seen to be the natural consequence of a new earth layer overlying an earlier one; and the percolating rain waters become recognized as the agent which dissolves the elements in the new layer; these elements then impregnate and crystallize with the elements of the lower layers.

II

PRE-HISTORIC FOREST FLOORS

IN PRIMEVAL FORESTS there was formed a deep bedding of vegetable matter which constituted their floor or ground—such as is noticeable to a lesser extent during a walk through woodlands today. This floor consists of the remnants of tree trunks, twigs, leaves, and any buds, blossoms or fruits thrown off in the yearly life cycles of plant life. A topsoil is thus developed, and this annual growth continues throughout an epoch of time lasting between 6,000 and 7,000 years, resulting in a considerable thickness or depth of newly developed earth materials.

If a raging fire consumes a forest, a new forest will eventually grow up in the same area. The layer of ash will appear as a tell-tale streak or vein in the rock layer into which the dead vegetation becomes metamorphosed during susequent epochs.

During a normal careen of the globe, as described in Part One of this book, a layer of sand, shells, and other flood debris from the oceans may also be laid over this former forest floor in varying depths. And this area of the earth—with its topsoil formed during an entire epoch—will be moved by the careen of the globe to a new latitude, where new forests will grow, either temperate or tropical, depending upon the new location.

Regardless of the latitude to which the area is shifted, or what debris covers the area, the usual secondary forces of nature continue at work—including rains which fall and sink into the ground. Water, under certain conditions, as it percolates through the new top layers of the earth takes up, or dissolves, silica and other chemical elements in the sands and shells of the newly overlaid

Pre-Historic Forest Floors

upper stratum. Now containing minerals in solution, the waters slowly penetrate the vegetable matter of the organic substrata, causing petrification of the vegetation.

The minerals are given up by the percolating waters. The vegetation—being in an amorphous condition—produces a different effect of petrification from that found as trees or produced from vegetation having definite solid shapes. The dissolved minerals crystallize according to their natures and combine the chemical elements of the amorphous vegetation with their own elements, resulting in the crystalline structures recognized today in granite and similar rocks, now normally classed as igneous rock formations—that is, rocks formed by volcanic work.

As an example of the work of ancient causes which produce the same effects even today, let us consider the creation of the fossil forests of Yellowstone Park, described in Part One. In the *U.S. Geological Survey, 1912, Trees and Vegetation*, Arnold Hague mentions about 2,000 springs in the Yellowstone area, mostly alkaline hot water springs, and states that ⅓ to ⅔ of the mineral content of the waters is silica.

When the hot spring waters sink into the ground, become chilled while percolating amid the buried vegetable ground materials, and have to give up their minerals, it is found that most of the minerals have become captured by the vegetable materials, or have captured the vegetation with which the minerals have crystallized. Thus, the petrification of trees and other smaller vegetation is now going on underground.

Several hundred thousand years have evidently been required to produce the 27 non-contiguous strata of forest tree levels in Amethyst Mountain, along the Lamar River (page 36). The silica became quartz, opal, and granite rock.

When it is hot at 194° F.—as in hot springs or geysers—water will absorb and carry more than twice the amount of silica absorbed when it is cold at 68° F. and is percolating in the ground. Conversely, if water is fully saturated with silica when hot, it must surrender more than half its load of silica when it is chilled to 68° F.

"The remarkable efficiency of silicification processes at least indicates that silica can be transported in solutions of such mobility and tenuousness as to effect widespread penetrations and impregnations." (C.S. Hitchins, in *Bulletin of Institute of Mining and Metallurgy*, Vol. XLIV, 1935.)

We have shown how soils build up through vegetation and that trees become petrified when underground. It follows that plants, grasses, and other particles of the soil also become petrified. Salient examples are the glass-like obsidian strata and the immense quantities of stone-like, glassy materials of angular shapes, or breccia, which have resulted from the weathering and decay of these fossilized materials when exposed on the ground in Yellowstone Park.

The rock structures in which these trees and other vegetation are entombed also contain great numbers of impressions of plants —many of which are perfectly preserved, including roots, stems, branches and fruiting organs, as well as grasses. (Knowlton, in the *U.S. Geological Survey, 1944, Yellowstone.*)

These granitic rocks appear to be "soil rocks", such as have once been soil. The identified vegetation could not have developed and withstood the molten lava of the laccolith theory or the fine volcanic dust of another theory. To say that the rocks are "igneous" is now superfluous and unnecessary. It seems more reasonable to account for these so-called "igneous" rocks by the new theory of "soil rocks," and the word "orgneous" would more correctly express their nature.

More detailed data linking vegetation and granitic rocks are given later (pages 248 and 249). For the moment, let it merely be pointed out that more than 25 of the mineral elements that occur in vegetation are also found in granites.

The theory that granitic rocks are igneous is attributed to the Scot James Hutton (born 1726). He saw that granitic rock layers were interleaved between other layers of rocks, and concluded that what he saw could only have been created by molten magma from the earth's interior. In Hutton's days nothing was known about Antarctica nor about the wobble of the earth.

The careenings of the globe—with the attendant growth of successive earth strata—explain the development of granitic rocks, and the occurrence of granitic rocks is offered as proof of the recurrent careenings of the globe.

During a world flood a layer of ocean sand together with other materials is assumed to have been thrown up and to have covered a forest of broken tree-trunks. Rain waters, having penetrated the sand layer, picked up silica and other minerals in tenuous solutions, carried them along in penetrations of buried trees and forest floors—causing petrification of both by replacement, impregnation and crystallization.

Quartz seams appear in many rocks. Silica (SiO_2) in tenuous water solutions appears to have crystallized into quartz. Clearly, if silica-laden waters were penetrating buried forest areas and became crystallized into quartz in fractures and cracks, then the activating solutions must have penetrated the vegetable and mineral elements of the soil from which the granitic rocks were eventually crystallized.

Thus, we have a new twentieth-century theory for the formation of granitic rocks. They are products of underground petrification. Petrification must be assumed to have taken place on the forest floors or the ground, which consists largely of vegetable matter, because it is generally so assumed in regard to the trees. If the latter is true, why not the former?

"Origin of Granite," Geological Society of America, Memoir 28, 1948, contains the following items, with my comments added: A.F. Buddington (page 36) finds two layers of granite separated by one layer of amphibolite. "The granites locally contain these interleaves of limestone." COMMENT: This shows that the area of the granite has been a sea bottom and an upland area, in previous epochs, in sequence. (Page 29) "The sillimantic granite . . . occurs almost wholly as fibres and fibrous aggregates in association with quartz." COMMENT: This shows that it was a forest floor which was metamorphosed into granite.

Frank F. Grant in the same memoir finds very little massive granite (page 47). On page 51 he mentions Duluth gabbro, of which the upper 10% is granite with somewhat irregular distribution. G.E. Goodspeed (page 68) finds that the dominant mechanism is the gradual penetration of activating solutions.

In *Internal Constitution of the Earth*, Chap. IV, page 8, L.H. Adams states: "Granites are hard and soft materials ... highly compressible and less compressible mineral grains." COMMENT: If this condition does not apply to igneous rocks, then volcanic lavas and granites are different types of rocks.

If all of the so-called igneous rocks were once in the same molten mass, seething into soupy form at excessively high temperatures, after which they cooled and crystallized, how can we account for hard and soft materials remaining separated in the same rock? We cannot account for the many differences in the so-called igneous rocks when we suppose them all to have been ejected from the same cauldron!

When the modern theory of volcanoes (see page 236) becomes accepted and replaces the prevailing "molten internal core of the earth" theory, granitic rocks will no longer be classified as "igneous." Volcanic rocks generally contain less silica than granites (see page 249). Sandstones and limestones as well as granitic rocks become crystalline volcanic lava when fused by excessive earth electric currents; but the fact that granites are crystalline formations does not prove them necessarily to have been lava magma.

Unlike ingots cast from molten metals, which have uniform internal structure, granites are not uniform internally. They appear to resemble dirt and debris in tiny horizons which have become crystallized. They split apart on their bedding layers when forcibly broken. Many are classified by their bedding characteristics.

Gneisses are a laminated or foliated group. The orthoclase group consists of granite rocks that break in directions at right angles. The plagioclase group includes those having oblique slanting cleavage or splits. Pyroxene—abundant in some granites—is often laminated. The word itself comes from the Greek, meaning fire stranger, which connotes "not igneous."

Photomicrographs, showing cross sections of various types of granitic rocks, reveal their contents. Some are mosaics, and the particles in the mosaic show bedding layers, indicating that the

Pre-Historic Forest Floors 231

mosaic particles were formed separately, by sedimentation, before they were combined into the present granite rock.

Petrified vegetation, as seen in micrographs (see U.S. Geological Survey, Monograph 32, Part 2, 1899, *Geology of Yellowstone National Park*), appears to range, in varying amounts, from grasses and wood dust to small vegetable fibers, with or without an occasional grain of dirt. "Orgneous" would appear to be a more accurate definition of these rocks than "igneous."

The petrified trees of Amethyst Mountain, Yellowstone Park, are composed mainly of silica; those at Cairo, Egypt, are composed mainly of calcium. The transportation of a knocked-down and buried forest to a warm climate or to an undersea location—due to a normal careen of the globe—may be considered a prerequisite for the calcification of the wood. Wood which has turned to stone (excluding coal) is generally found to be silicified, calcified, or partly silicified and partly calcified. The colorings of the petrified wood are due to other chemical elements and vary from red to blue, from garnet to turquoise.

When a soil layer is superimposed on another soil layer after a flood, a formation develops which is different from a superimposed silica layer. In the tropics—when mining tin ore—normally hard rocks, like granites, are found to be so soft that they can be dug with a spade to depths of scores or even hundreds of feet. R.H. Rastall, in *The Encyclopaedia Britannica*, 1947, Vol.X, pages 158-59, describes this and explains it as decomposition of rocks due to chemical action. If the theory of orgneous rocks is accepted, these soft granites are now identified as partially crystallized vegetation.

Another example of non-silica type of rock formation is given by Nordenskiöld in his description of frozen Wood Hill of the New Siberian Islands. In his *The Voyage of the Vega*, he states that the tiered rock strata showed layers alternating between bare rock strata and strata containing fissile bituminous tree stems. The trees were not petrified but had turned into coal. In Yellowstone Park the trees have turned into stone.

Coal beds are found below layers of granite. But if granitic rocks were of igneous origin and were ejected from a molten magma of the earth's interior, then the heat of the magma would have destroyed the coal layer when it penetrated through the volatile coal!

The Coal Resources Section of the U.S. Geological Survey advises: "At several places in Wyoming large masses of granite overlie sedimentary rocks of Tertiary age, which may contain coal at depth. The granite is not in contact with the coal. The occurrences—one on the south side of Owl Creek Mountain, and another on the southwest side of the Wind River Mountain— are interpreted by most geologists as being the result of thrust faulting that took place when the mountains were uplifted."

Coal beds are found below layers of limestone. Limestone was created in the oceans by corals and shellfish. Coal is formed from vegetation that grew in upland areas. Limestone is now found in the upper strata and coal beds in the lower strata of the earth. These natural phenomena cannot be accounted for except by the theory of successive creation of the earth's strata, during the epochs of time between the recurrent careenings of the globe.

The granitic rocks of Yellowstone Park are identified as soil rocks on account of the vegetable growth contained in them. During the very same epochs of time, ocean rocks were also being created—including limestones, marbles, and sandstones. Where land and water met, different types of rocks were developed— such as clays, shales, and slates.

The formation of soil rocks and sandstones is due to the fact that water absorbs, dissolves, and carries minerals in solution and in suspension, and that minerals form into crystals when changing from aqueous to solid form.

The mineral content of drinking water is indicated by the scales which form on the bottoms of kettles used for boiling water, and also by the tiny crystals or flakes which are the sediment from the microscopic mist produced by cold water spray-type home humidifiers and appear as a dust on floors and furnishings.

Mud—which is a viscid form of silt—is wet dirt and is com-

Pre-Historic Forest Floors 233

posed mostly of tiny, even microscopic, mineral and vegetable particles. On drying out, the absorbed minerals crystallize according to their natures and solidify with the organic and inorganic particles. Mud changes to hardpan. Hardpan changes to rock.

In penetrating strata of vegetable matter, the water-borne minerals have replaced the vegetable matter in petrified trees and wood, and have crystallized to unite closely and intimately all of the molecules of the amorphous vegetation and join them so as to form an integrated whole.

In penetrating sand strata, the water-borne minerals have permeated and impregnated the interstices between the adjacent grains of sand—welding them together to form solid sandstones.

Embryonic rocks, which have not yet become fully solidified, are common among sandstones and granitic rocks. Their powers of cohesion vary over a very wide range under test pressures. Some of these soft rocks—such as those found in the Long Island terminal moraine—resemble the mottled gray granites common in the vicinity of New York City.

The softer stones will crumble under finger pressures and are commonly called rottenstone. Some granitic-type samples crumble with a tendency to break in parallel planes of about a sixteenth of an inch in thickness. They are evidently not deep-layered rocks which were once hard but have now softened: but they appear to be recently formed rudimentary stones—still undergoing the process of formation, but not yet fully solidified.

The prevailing notion that rocks do not grow is refuted by recent developments with regard to mica—an element that is present in many granitic rocks. Today mica is being grown from what are called tiny seeds of mica, and the synthetic mica has the same properties as the natural mica of granitic rocks. (See Technical Report 1040, National Bureau of Standards.)

Rock making, by the injection of silica and calcium, is today an established commercial routine. Engineering specialists have developed techniques for the strengthening of the supporting soils of building foundations, highways, and mines. Subsoil earths of less firm textures are converted into rock by the injection of silicate of soda, calcium chloride, and other activating agents, which gel or crystallize with the soil.

Heat cycles have been important in developing some of the existing types of rock. Water, when warm, absorbs and holds double the amount of silica that it can continue to hold when cold—as already indicated. Magnesium and other minerals appear in dissolved form in warm water; but calcium is dissolved in cold water but is rejected when the water gets hot.

Cold water—issuing from underground springs in Italy and Spain, for example—is found to contain calcium, dissolved from underground limestone. When these cold spring waters become warmed up in the sun, in slow-flowing streams, the calcium crystallizes out as travertine and as aragonite—two forms of very hard, crystalline rock.

Thus, in an ever continuing process of creation, rocks develop as the result of two opposite heat cycles. Granitic rocks are formed in Yellowstone Park and elsewhere because water absorbs silica and other minerals when hot and rejects them when it gets cold. Travertine and aragonite are formed because water, containing carbon dioxide, absorbs calcium when cold but rejects it when warm.

Topsoil is well known to all dwellers in temperate zones. It is dirt—consisting mainly of decayed vegetation, including roots and wind-blown dust grains. Under certain conditions soil becomes hardpan and hardpan becomes rock.

The creation of certain kinds of dirt created from vegetables can be traced to very humble beginnings. The lichens, for example, found in Antarctica and in Greenland are described by Rutherford Platt as "A strange partnership of fungus and algae, whose acid-bearing growths hold fast to glasshard boulders. Lichens have no need for soil, but producing it, lay the cornerstones for flowers and trees. They are the plant-world pioneers, bringing life where none existed." (*National Geographic Magazine*, April, 1957).

The Creation of Limestone

THE SEQUENCE from grasses to corals to limestone epitomizes the creation process. Limestone is calcium carbonate ($CaCO_3$). It is developed mainly from corals, shellfish, and the various crustacea and foraminifera. The calcium is created in vegetation. To illustrate, calcium is the bone-building material in the bodies of calves, and very obviously it is produced from the grasses which cows eat. Grasses are rich in calcium. From these well-known facts the theory develops naturally that the calcium contained in limestone comes from vegetation.

Billions of tons of grasses are grown each year. When dead and decayed they are leached by fresh rain water which picks up calcium and carbon and carries these elements to the oceans by way of the rivers.

Many references to the percentages of calcium in grasses, legumes, and other vegetation are condensed and assembled in "Bibliography of the Literature on the Minor Elements and their Relation to Plant and Animal Nutrition," by the Chilean Nitrate Educational Bureau.

Research conducted at Oklahoma State University on chemical elements of the earth found in vegetation is reported on by H. A. Daniels in *The Journal of the American Society of Agronomy*, 26, 1934, No.6, pp. 496-503. He reports: "368 samples of 25 different kinds of grasses showed an average calcium content of 0.351%. 336 samples of 12 kinds of legumes showed an average calcium content of 1.373%. Grasses of India showed 0.314 to 0.814% calcium. Bermuda grass was 0.659% and alfalfa was 1.507% calcium. A listing of legumes showed 1.040 to 1.833% calcium."

In his report Daniels further states, "Crops high in mineral content but low in calcium and phosphorus remained low when grown on fertile soil. Crops high in calcium and phosphorus always had large amounts, even on poor soil." This can be paraphrased by saying that the percentage of calcium in a plant is not a function of the soil in which the plant grew; but it may be influenced by it.

All of the many successive layers of limestone rock grew originally at the bottoms of shallow seas in warm climates. They are now observable where mountains have been elevated or canyons gouged out. Each of the individual limestone strata of the earth grew in one of the successive epochs of time which ended many thousands of years before the epoch of time in which we live. The calcium which created these stone layers was absorbed from sea waters by animal life and it was constantly replaced through the agency of vegetable life. That same creative process of metamorphosis is going on today.

Much of the calcium now being carried to the oceans by the rivers has been dissolved and taken from limestone rocks by flowing fresh water. It is being used over again by animal life to create new coral polyps and new oyster shells in tropical and temperate zones respectively.

Volcanoes and Hot Springs

VOLCANOES FUNCTION somewhat on the same principle as that on which the electric arc furnace is based, where electric currents melt iron to be used in certain types of steel. Hot springs develop on the principle of the electric blanket that keeps you warm on cold wintry nights.

Very small electric currents flow through the wires of the electric blanket, creating a gentle heat. Excessive electric currents would melt the wires. The electric earth currents causing hot springs are not strong enough to melt the rocks through which they flow, but volcanoes come about when the electric currents are strong enough to melt the rocks. The elimination of the older hypotheses of a molten core of the earth and a shrinking crust leaves electric earth currents as the more rational explanation of the concentrations of heat which cause volcanoes and hot springs.

The internal heat of the earth—aside from that received directly from the sun—can be readily accounted for as being a condition related to the surface of the earth and caused by myriads of electric currents flowing in the upper earth strata. We

Pre-Historic Forest Floors

know that all electric earth currents heat the materials through which they flow.

Measurements of the internal heat of the earth at different places show great and conflicting variations. They have been reported rather fully in *The Internal Constitution of the Earth* published by The National Research Council. Those data and others disprove dramatically the presently accepted theory that temperatures below the earth's surface increase uniformly with depth.

The mistaken assumption that the earth's internal temperature increases with depth was underscored when the 12¼ mile-long Simplon Tunnel—connecting Switzerland and Italy—was built during the first decade of this century. The records show that great fear was expressed at the time that the tunnel temperatures would be abnormally high.

The level of the tunnel was, therefore, made higher than the grade of the railroad required, in order to keep away from the assumed "hot core" of the earth, it being generally assumed that the temperature gradient for the internal heat of the earth followed the surface configuration. The hot internal temperatures which had been predicted were found not to exist. Today—as shown by a recent test—the temperature in the middle of the tunnel varies between 75 and 84 degrees Fahrenheit.

Proofs that the earth is not hot at great depths are also provided by deep-water wells and by oil and gas wells, after the heat of the drillings has been dissipated. We find ice-cold waters in the abysmal deeps of the oceans—nearly seven miles nearer to the assumed "hot core" of the earth than to the surface waters. The deep water does not get hot and no convection currents of rising hot water have been found. There is little or no evidence of the conduction of heat through the ocean bottoms to the ocean waters (*The Oceans*, by H.U. Sverdrup, pages 110 and 738). All this refutes the current theory to the effect that the earth's "crust" is very thin under the oceans, and also the theory holding that there is a hot molten core below it. The empirical data contradict all those theories. Electrically created heat is the main cause of the variations of the internal heat of the earth at different cross sections, and at different times for the same sections. This can be checked by research in many cases. Electric voltages, from thousands of causes, create the earth electric currents, and the earth currents are the cause of the heat. Local voltages are equalized by the flow of the electric currents.

A knowledge of electric earth currents is a prerequisite for comprehending the electrical origin of volcanoes and hot springs and the nature of electric voltages must be understood for one to understand electric earth currents.

An electric voltage or potential was called an electromotive force in the early days when batteries were the prime source of generating electricity commercially. These batteries consisted of two dissimilar metals, usually immersed in an acid solution. It was found that the metals generated different electrical potentials between themselves. Later it was discovered that voltages or differences in potential could be created by moving a copper conductor in a magnetic field so that the conductor cut across the magnetic lines of force or flux.

Electric voltages in the earth are today being created, like battery voltages, by two dissimilar metals, and, as in the case of dynamo voltages, by externally driven conductors of earth material being moved across magnetic fields of force.

It was also discovered that whenever an electric voltage or potential was created and an electric conducting medium was used to make an electric circuit, by connecting the positive to the negative terminals, electric current would flow and a contact maker and breaker, or electric switch, was needed in the electric circuit either to permit or shut off the flow of electric current by closing or opening the switch. When electric current flows from one terminal to the other of a dynamo, the voltage is sustained by the driving force of the prime mover. In a thunder cloud, by contrast, the voltage is dissipated and equalized by the flow of electric current in the flash of lightning.

It was discovered that the rate of flow of an electric current was directly in proportion to the voltage and inversely in proportion to the resistance of the electric circuit. This is expressed as Ohm's Law. The formula is

$C = \dfrac{E}{R}$ where C is current, E is voltage, and R is resistance.

It is usually expressed in amperes, volts and ohms, respectively.

Then it was found by test measurements that the flow of electric currents heats the mediums through which they flow in proportion to the square of the electric current multiplied by the resistance of the material through which it flows. The heat is measured in watts of energy. The law is written $W = C^2R$, where W is watts, C is current in amperes, and R is resistance in ohms.

One watt is the product of one ampere multiplied by one volt and is equivalent in energy to 0.0009477 B.T.U's (British Thermal Units) per second, also approximately 3.4 B.T.U's per hour and 82 B.T.U's per day of 24 hours.

Data are lacking on the volume of the earth currents which cause fusions of the upper strata of mountains; but the rate of current flow, measured in amperes, can be readily approximated from the above formula if the accumulated B.T.U's of heat energy and the elapsed time of the heat build-up can be computed, together with the cross section of the fused area and the distance and direction of the current flow.

Volcanic craters often occur at or near mountain tops. This proves to be entirely natural when the cross sections of the mountains are analyzed. All earth strata, when originally created, were laid down in approximately horizontal layers. Mountains are upthrusts which bend and crack the upper strata. A rounded or circular-type mountain may be visualized as half an onion, resting on its cut section, with some of the peak layers of the onion's successively created skin layers damaged or cracked by folding or faulting.

Earth electric currents, traveling in the rock strata, become concentrated at the peak, where they are retarded by the high resistances of the fractured sections of stone; or the higher voltages of the ambient air, due to elevation, may influence the flow of electric currents. The voltages of the air vary greatly with location, weather conditions, and time of day, the voltage gradient per foot of elevation ranging from around 10 to 80 volts. (See *Physics of the Air*, by W.J. Humphreys, page 398.) Air voltages at the tops of mountains are correspondingly high. St. Elmo's Fire at the tips of ships' masts is a visual display of the electric current always flowing between air and ground and caused by the differences in the voltages of air and ground.

It becomes easy to conceive of a volcano resulting, when voltages become too high and there is an uncontrolled excess of electric current. The fault in the rocks becomes overheated by the electric current because of its high resistance. A blowout occurs in the electric circuit—just as blowouts occur in any other electric circuits where a loose joint or fault occurs.

The fusing of the rock materials is nature's way of repairing the break in the electric circuit. The volcanic eruption is the incidental display of the process and nature of the repair job.

The variations in the chemical compositions of lavas from different volcanoes indicate that magma basins are isolated and hence imply that they are enclosed in solid rock and are not directly connected with a terrestrially large and general source such as the hypothetical hot core of the earth.

Some specific proofs follow:

Paracutin volcano, our youngest mountain, was born in 1943. Its appearance as a volcano is most reasonably and simply explained as being the result of earth electric currents which overheated the earth substrata at that particular location. The farmland got hot.

Finally, a fusing temperature was reached. The earth materials became incandescent and soft. They lost their rigidity and resistance to the compression pressures which exist in the upper earth strata, such as the pressures that cause excavations to cave in unless they are shored up and buttressed against the lateral pressures by bulkheads.

Earth pressures at the Panama Canal force the bed of the canal upward in locations where the rock is soft. Coastal mountains and rocks thrust up in Antarctica and Greenland are the results of the inland pressures of the central, growing ice masses.

Pre-Historic Forest Floors

The U.S. Navy has sponsored research on earth pressures, since they affect their piers and bulkheads.

At Paracutin these lateral pressures caused earth materials to flow sideways, replacing the soft molten earth materials; these became elevated by the lateral pressures into the shape of a dome or blown-up bladder—to the height of about 820 feet above the plain of the former Mexican farmland.

Paracutin is a new mountain but an old phenomenon. Hundreds of other great cinder cones in the area can be rationally attributed to the internal heat of the earth, generated by the development of electric voltages which created earth electric currents which became too concentrated and strong for the rocks to carry without becoming overheated.

The last previous eruption in this area was that of Jorullo mountain—about 80 to 90 miles to the northwest. This mountain, was created in 1759 and became inactive in 1760. Both mountains are less than a hundred miles from the Pacific coast and at elevations of about 7,400 and 4,300 feet above sea level, respectively.

If magnetic types of rock, magnetite, can be discovered in the vicinity of the volcanoes, where the earth electric currents flowed, they will be found to possess a magnetic saturation many times greater than the magnetisms being imparted to rocks by normal earth electric currents. Verification of this fact by future research is predicted—this will, in turn, confirm that earth electric currents have caused those volcanoes.

Electric storms frequently occur in the earth's atmosphere, often accompanied by a display of aurora borealis. Such storms interrupt grounded telegraph and telephone systems. These interruptions are associated with an overcharged condition of the ambient air.

Electric currents and magnetism are always inseparable; therefore, the overcharged ambient atmosphere creates a voltage condition of higher potential in the conducting earth materials, rotating from west to east below. The conducting sections of earth materials cut across magnetic lines of force and, like the windings on a dynamo armature, create higher potentials. Higher potentials with unchanged earth resistance create an increase in the rate of flow of the electric earth currents in the circuits of the earth's materials.

Electric storm voltages have been recorded as high as 500 volts, and the resulting electric earth currents, in grounded telegraph circuits, have caused sufficient heat to burn out coils and char paper insulation on cables. This gives us an analogy for the earth electric currents which blow the tops off mountains, cause volcanic mountains, and create hot springs.

Thousands of forces are at work creating earth electric currents by means of changes in the electric voltages occurring in the upper strata of the earth and in its ambient atmosphere.

Some of the causes of the changes in electric voltage conditions are heat, friction, and contact of dissimilar metals or materials, including changes in voltage due to changes in temperature, as in thermo couples and thermo piles.

Other causes of electrical potentials are impact, percussion, vibration, disruptive cleavage as in tearing paper, crystallization, combustion, evaporation, pressure, pyro and piezo electricity in crystals, animal and vegetable electricity, and chemical, electrochemical and dynamo electricity resulting from voltages being created by power-driven conductors being moved across a magnetic field of force, or magnetic fields of force moving across conductors, including earth strata.

The rock strata of conducting materials of the earth, moving with its daily rotation, cut across the lines of force established by the incoming energy from celestial space, including the energy of the sun's light and heat radiations, which creep from east to west against the rotation of the globe. The rock strata resemble the windings of a big, externally driven dynamo, creating earth voltages.

It is recognized at once that the voltages thus created must cause electric earth currents to flow. These voltages are mainly surface voltages, and the currents flow in inverse ratio to the resistances of the earth materials.

The electromagnetic force of the sun's rays is demonstrated by the sun flower with the single broad flower and stalk, which

Pre-Historic Forest Floors 243

turns with the sun from morning to night, and then at sunrise next day, reverses itself and again follows the sun on the succeding day. James Clerk Maxwell formulated the rule that every portion of a circuit is acted upon by a magnetic force urging it in such a direction as to make it embrace the greatest possible number of lines of force. It was from that principle that the direct-current electric motor was developed.

Gravitation—based on the new theory of tremendous amounts of electrical energy reaching the earth continuously from the suns of celestial space—fits the theory of earth electric currents like a piece in a jigsaw puzzle. The theory holds that energy rays from celestial space, shot out by countless billions of stars, collide with the earth, create its weight, and by striking unevenly make it rotate.

This incoming electrical energy, together with the sun's light and heat radiations, acting, for example, on thermo couples and thermo piles of adjacent dissimilar materials, produce spontaneous voltages which cause the flow of electric earth currents; many of these change in intensity between noon and midnight, since they are tied in with the rotation of the earth on its Axis of Figure.

Earth electric currents do not flow uniformly but vary with the conductivity of the earth's strata. Some are good and others are poor electrical conductors. Some are dielectrics when dry, but become good conductors when wet. Sandstone, marble, and limestone, which are poor conductors when dry, become fairly good conductors when wet.

Earth electric currents will become concentrated in the good conducting earth materials. Activities of volcanoes should be expected and predicted when the materials of the earth in that particular area are wet or moist and become good conductors of electric currents.

Telltale magnetic rocks whose directional pointings differ from the directions in which the earth's electrical field is now magnetizing similar rocks, are described in Part One (page 85). The failure of present earth electric currents to change the orientation of some of these old magnetic rocks supports the theory that the earth electric currents are generally near the surface—but not always.

Earth electric currents are found to be present at considerable depths in some areas. For example, a constant volume of steam arises from a 3,189-foot deep drilling in Geothermal Valley near Rotorua, North Island, New Zealand, indicating the presence of deeply flowing earth electric currents; and similar thermal conditions exist in many other regions, including Iceland, Tuscany, and Yellowstone Park.

The universal presence of earth electric currents and their variations in magnitude and directions support the theory that they are the causes of volcanoes and hot springs.

Two erroneous but widely held theories hold that the earth has a molten core and that the core is largely of iron, a fact which causes the inductor compass to point north-south. Iron loses its magnetic characteristics at red heat. When molten, iron is nonmagnetic and, therefore, a molten iron core of the earth could not affect the pointing of an inductor compass.

The Inductor Compass

THE INDUCTOR compass, commonly called the magnetic compass, reveals the presence of earth electric currents everywhere. The compass action indicates that it responds more nearly to the laws of electromagnetic inductance than to the laws of magnets.

Electromagnetic induction results from the myriad and all-pervading electric earth currents. The compass needle turns and dips with a magnetic moment of force, but it does not move itself toward a magnetic pole—the way a magnet does. In most places on the earth's surface it does not point directly to the earth's magnetic poles.

If fastened to a block of wood, freely floating in water, it will turn and point in a north-south direction, but it will not move toward either magnetic pole. It does not move toward a magnet the way a toy magnetic fish, floating in a bowl of water, moves toward a magnet. It has a north-seeking and south-seeking end and is held in its north-south pointing position by

Pre-Historic Forest Floors 245

the magnetic flux established by the electric earth currents. The inductor needle can be used to measure the intensity of the earth's magnetism at any location. These measurements, in turn, give the direction and intensity of the coexisting activating earth electric currents.

No earth current exists without creating a magnetic field of force. If one is created, the other one simultaneously exists at right angles to it. The two always coexist.

Observations of the compass and dipping needle at many places on the earth's surface show daily and annual variations in direction and magnitude, which appear to be a natural result caused by the variations in the local earth currents and their coexisting magnetic fields of force. An iron core in the earth's center, even if cold and not molten, would not produce these observed variations.

Observations of the variations in the declination of the compass for ship routes all over the globe, show that the pointing directions follow an irregular curved line—not a straight nor a meridian earth circle line. This would indicate that the chief motivating force of the inductor compass must come from local earth magnetic fields of force, created by the coexisting earth electric currents. It disproves the theory of a unidirectional great earth magnet.

Earth electric currents, if flowing from east to west, would cause an inductor compass needle placed above the currents to point north; if placed below the currents, the needle would point south. This is based on the observed action of the compass needle when held above and below a wire carrying a unidirectional current.

A simple test of this would be to lower an inductor compass into a mine shaft, so that it is placed below the earth's surface currents. Theoretically, the needle should reverse its direction of pointing. Engineers consulted for corroborative empirical data say that they never use the compass in mines because it is unreliable—due to the electric currents in adjacent wires, including trolley wires, and to the nearby presence of steel rails and iron and steel pipes in the mines.

Aviators have reported that the inductor compass becomes useless and gyrates wildly upon entering thunderstorm areas, where electric currents are rushing about in all directions between clouds. In this case it is clear that it is electric currents, not a magnetic core of the earth, that control the needle fluctuations.

The knowledge of earth electric currents, gained through observations of the actions of the inductor compass, shows their all-pervading existence and adds to the proofs that make it reasonable to assign the causes of volcanoes and hot springs to their excessive concentrations in specific areas.

The first reference to an iron core in the earth was made by William Gilbert in 1600. He called the earth a great magnet with an iron core. Later, the nebular hypothesis—expanded mathematically by LaPlace—was based on the assumption that the heavier elements would be concentrated in the center of the earth; and iron is a heavy metal. The current edition of *The Encyclopaedia Britannica* carries adequate refutation of LaPlace's mathematics as applied to the earth.

Among former attempts to explain the creation of the earth, within the scope of the laws of nature, are the nebular hypothesis, the planetesimal hypothesis, and the tidal theory. All leave unexplained why the earth has been created in strata and how and why fossil plants and animals got into the lower layers.

The new cosmic geophysical theory here presented and explaining the origin of the earth's materials also explains why the earth is composed of stratified layers, why well cores, drillings, excavations, cliffsides, and ravaged mountainsides show earth formations by periods of time, and why clays, shales, and slates and some other rocks show their development and growth by years, like the annual rings in trees.

This new theory of the continuous creation of earth materials from radiant energy—resulting in rock layers which generally represent epochs of time between the recurrent careens of the globe—is a contradiction of the theory which presently prevails in academic circles that the so-called igneous rocks are the ultimate source of all of the materials of the earth's upper layers

Pre-Historic Forest Floors

(excepting limestones, peats, coal, fossil trees), and that they have become solidified from a previous molten condition, forming the so-called crust of the earth.

Contradictions of widely-held popular theories are supported by the following evidence:

> The earth does not have a molten core
> Laccoliths were not produced by molten rock extruded from below
> The earth does not have a shrinking crust
> The earth is not shrinking in diameter

These evidences also show that:

> The earth has been constructed in consecutive layers of strata
> Fossils in the successive layers indicate the ages of the strata

It must be recognized that all academic geology is not a body of indisputable and immutable truth. Geology has never been anything more than a body of well-supported probable opinion! The academic treatment of geological phenomena must eventually concentrate on the opinions requiring the fewest postulates to explain them—a well accepted philosophic principle. The postulation of earth electric currents is such a simple and unifying theory that explains a host of geological phenomena.

Minerals in Vegetation

THE CHEMICAL elements that constitute the average content of all vegetation can be assumed to be $C_6H_{12}O_6$ with about 1% mineral matter; or 6 atoms of carbon, 12 of hydrogen and 6 of oxygen, plus minerals. This information was kindly furnished in a communication from Dr. Richard M. Klein, Curator of Plant Physiology of The New York Botanical Garden; and his estimate was confirmed by the Director of the Division of Radiation and Organisms of the Smithsonian Institution.

Half a century ago wood and vegetation were generally considered as being composed of only three chemical elements: carbon, hydrogen, and oxygen, and they were classed as carbohydrates. Today, vegetation is known to contain 54 of the 92 basic chemicals of the earth; and many of the minerals found in vegetable growths are currently considered as being essential to these growths—either activating growth directly or functioning as catalysts to promote growth, as when sprayed or painted on plants or embedded in the adjacent soil.

In "Bibliography of the Literature on the Minor Elements and their Relation to Plant and Animal Nutrition," published by Chilean Nitrate Company, thousands of references are given to books and essays pointing out the existence of the following chemical elements in vegetation:

Aluminum	Antimony	Arsenic	Barium	Beryllium
Boron	Bromine	Cadmium	Calcium	Ceasium
Chlorine	Chromium	Cobalt	Copper	Fluorine
Gallium	Germanium	Gold	Iodine	Iron
Lead	Lithium	Magnesium	Manganese	Mercury
Molybdenum	Nickel	Platinum	Radium	Rhodium
Rubidium	Selenium	Silicon	Silver	Sodium
Strontium	Sulphur	Tellurium	Thallium	Thorium
Tin	Titanium	Uranium	Vanadium	Zinc

(Nine rare chemical elements have also been identified in plants, but have not been listed here.)

The above-named chemical elements—found in plants and in the animals which eat the plants—all become normal constituents of the soil, because all animal and plant life at death become part of the soil. There is, therefore, a constant buildup of earth materials containing the chemical elements found in plants and animals. The chemical elements found in plants are also found in the rocks. More have been found in rocks than in plants; but it also happens that data on rock analyses have been gathered more extensively by more persons and for a longer period of time than have data on the mineral analyses of plants.

Chemical elements identified in some of the common rocks are listed below. Seventy-five of the chemical elements have been identified in rocks, and 54 in vegetation.

CHEMICAL COMPOSITION OF ROCKS*
(Percentages of total weight)

	I Igneous	II Sandstone	III Limestone	IV Shale	V Volcanic
SiO_2	59.12	78.31	5.18	58.11	47.60
Al_2O_3	15.34	4.76	.81	15.40	6.01
Fe_2O_3	3.08	1.08	54	4.02	3.17
FeO	3.80	.30		2.45	4.59
MgO	3.49	1.16	7.89	2.44	14.43
CaO	5.08	5.50	42.57	3.10	21.52
Na_2O	3.84	.45	.05	1.30	0.70
K_2O	3.13	1.32	.33	3.24	0.76
H_2O	1.15	1.32	.21	3.66	0.08
CO_2	0.102	5.04	41.54	2.63	
TiO_2	1.050	.25	.06	.65	1.52
ZrO_2	0.039				
P_2O_3	0.299	.08	.04	.17	
Ci	0.048	Tr.	.02		
F	0.030				
$(Ce,Y)_2O_3$	0.020				
Cr_2O_3	0.055				
V_2O_3	0.029				
MnO	0.124	Tr.	.05	Tr.	0.13
NiO	0.025	.05			
BaO	0.055			.05	
SrO	0.022				
Li_2O	0.007	Tr.	Tr.	Tr.	
Cu	0.010				
Zn	0.004				
Pb	0.002				
H_2O		.31		1.33	
SO_3		.07	.05	.65	
S			.09		
C				.80	

* These tables present averages of several hundred samples. Table I is from Committee of National Research Council; II, III, IV from *U. S. Geological Survey;* and V from *American Journal of Science.*

A Copper-Bearing Tree

... was reported by Professor G.B. Frankforter of the University of Minnesota in *Chemical News*, Vol. 79, 1899, page 44. He states that "bright copper colored powder is disseminated through the pores of the tree ... granular copper ... some granules formed flakes as large as 1½ millimeters diameter ... found to be pure metallic copper ... 1.2 milligrammes of copper to 100 grammes of wood ... distributed heart 0.8, half-way 1.86, near bark 3.95, average 1.2." These data indicate an increase in copper content with increase in age of the tree, an oak, whose age and weight are not given.

Assuming that this one oak tree produced a half ounce of copper, lived one hundred years and then died and rotted by slow combustion—oxidation—its mineral contents left in the soil would develop to a depth of about thirteen feet during a 6,000-year epoch of time (see page 36); assuming also that during that period of time many more such copper-bearing trees grew in that same area and disappeared at death, then that soil can be assumed to have copper particles disseminated throughout its mass. It would be classed as porphyry copper ore.

A forest fire destroying such copper-bearing trees in past ages could account for our now finding molten globules of copper in the former soil, which is now rock and displays an occasional dark streak of rock strata, which indicates that it was once a burnt-over forest floor. This résumé of developments indicates that such rocks were never igneous.

The recurrent roll-arounds of the earth explain many sporadic locations of mineral ores. The copper-bearing oak tree is a specific example of minerals occurring in a land plant (see page 247 for percentage of minerals in vegetation). It is further postulated that copper was developed in the lush vegetation of some ancient sea bottoms, and enveloped in successive layers of what was once the accumulating muck of thousands of years. In our present epoch of time some of these deposits happen to be high above sea level.

The high elevations of some porphyry copper deposits and

Pre-Historic Forest Floors

their large areas have an analogy in the elevations and comparable sizes of chalk deposits. The chalk cliffs of Dover, England, are now high above sea level and cover many square miles, including parts of the formation that is now across the English Channel, in Normandy. The materials of the chalk cliffs were created on the ocean bottoms by globergerina, during a period covering thousands of years. The animals themselves have disappeared, just as the oak trees that bore copper disappeared, but their mineral content has remained as a part of the growth of the earth.

In view of the recurrent roll-arounds of our globe, and of the successive epochs of time between earth revolutions, it is reasonable to postulate that copper and other minerals have at times been dissolved and then recrystallized with other elements of the earth, during epochs of time when they were at other latitudes, in other temperatures, and under conditions favoring such metamorphoses.

This new cosmic geophysical theory may also be the correct explanation for the occurrence of copper in large masses, as is the case in mines in Michigan. It is assumed that the copper grew in sea vegetation or in land plants.

Copper is among the minor or trace elements found to be invariably present to some extent in all plants and animals. I have tried to find seaweed, rich in copper, that would break loose and drift in ocean currents and settle in an eddy, hole or submarine canyon, and there disintegrate, leaving the copper and other minerals to accumulate as a residue, like ashes from a fire. At any one location it should be found in alternate layers of the earth's strata, and widely distributed over the earth. A likely source would seem to be the weed of the Sargasso Sea, known as sargassum, but research centers have so far no record of the copper content, if any, or of any minerals, in sargassum.

Seaweed research disclosed that titanium, a valuable and scarce metal, occurs in Scottish seaweeds, Fucas speratis, in concentrations 10,000 times stronger than that found in the enveloping waters. "Soaking a fresh Laminaria (seaweed) frond in water . . . does not remove the trace elements, which appear therefore, to be in insoluble form." (Black and Mitchell, in *The Journal of the Marine Biological Association,* Plymouth, England.) This statement would appear to be substantial proof that the trace elements do not arrive in the seaweeds from waterborne solutions; but rather that the trace elements in the water come from plant life which, in turn, develops it from the energy of light rays which become occluded or frozen as atoms making up the plant structures.

"Below about 300 feet depth, where the light rays fade out, marine plants cease to grow." (*Columbia Viking Desk Encyclopedia,* page 604). When the temperatures of the seaweeds are raised to incandescence or above, the atoms of the plants return to the state of radiant energy.

"Cadmium, chromium, cobalt and tin have been found in the ash of marine organisms and hence it is implied that they occur in seawater, although so far they have not been obtained directly." (*Lange's Handbook of Chemistry,* 1956, page 1107.)

Mineral ash may result from both vegetable and animal life. "The sea squirt, Phallusia mamillata, for example has 1,000,000 times more vanadium in its blood than the water it lives in; the deep blue blood of the octopus has 100,000 times as much copper," according to Professor Ernst Bayer of Tübingen University in Germany, who has studied the ability of marine animals to effect concentrations in their bodies of some of the rare metals found in sea water. (*Time Magazine,* May 15, 1964, page 90). Vegetation whose ash produces minerals is mentioned on page 235 for calcium and on page 250 for copper; on page 265 the creation of water from vegetation is outlined.

This creative process has been going on continuously—epoch after epoch—as part of the creation of the successive layers of the earth's minerals. If we know how and where metals are formed as minor trace elements, it will be easier to determine the likely locations of such metals in the ground.

Our increasing knowledge of the buildup of the materials of the earth, strata by strata, epoch by epoch, and of the successive changes in the latitude of the various areas of the earth's surface, affords an opportunity for introducing a new theory of the crea-

Pre-Historic Forest Floors 253

tion of some of the mineral ore deposits. The minerals of the earth have developed as explained elsewhere in this treatise, and ore bodies have grown *in situ*, always resting on the earth strata that were the surface of the earth or ocean bottom of the epoch of their creation. Shallow ore bodies are found to follow extensive synclines and anticlines of the underlying rock strata which supported them when developing, and the metals are generally found in veins, faults, sills, dikes, saddles, lenses, flats, sheets, reefs, and bedded formations, all of which are now at various angles to the horizontal.

Many metallic minerals are found along faults. Faults are likely to have occurred along lines of weakness and are naturally created along the unconformities between a former ocean bed and a new earth layer, because new earth material had not had time to become rigidly consolidated rock before the next rollaround of the globe occurred. Many of the metallic ores have been found in the ashes of seaweeds, and as the oceans cover about 71% of the earth's surface, it can be logically assumed that most of such minerals were created in the oceans. Many of the branching ore veins, now slanted or vertical, may have got this shape because the ocean bottom was not level, or because it had been grooved by the erosion of water streams when it was an upland area in the previous epoch.

Earth strata were generally created horizontally. Due to the repeated cataclysms, these once flat layers of the earth are now greatly twisted and distorted; most are now slanted, some are vertical and some completely overturned. There have been more than one hundred cataclysms, due to the recurrent roll-arounds of the earth, during the past million years, and that accounts for the twisted, jumbled condition of the inner rock strata.

The chemical elements classed as metals, which are most abundant in light spectra of stellar atmospheres as recorded by observatories, are generally also the most abundant in the top layers of the earth's strata. Also, there is considerable evidence that the composition of the chemical elements of the stellar atmospheres do not vary greatly from star to star, although spectra radiations do vary with the star temperatures.

Knowing that plants and animals grow at night, when the sun is not shining, as well as in the daytime, when it is shining (corn grows faster at night, the night blooming cereus opens only at night), we look around for the sources of energy for creating mass in the absence of sunshine; and we find it in the radiant energy rays sent to the earth, and absorbed by the earth's elements, from countless myriads of stars—each of which is a sun and a source of radiant energy rays.

It is well known that heat conditions vary on the surface of the earth, and that plants and animals vary accordingly. It is a truism to say that both plants and animals depend on the sun for heat and light. Animals and plants differ with the temperatures. There are torrid-zone, temperate-zone, and frigid-zone plants and animals—yet all feel the same drag of gravity exerted by the incoming radiations from celestial space. These radiations are identified in this treatise as being the major source of the energy from which the materials of the earth are created.

At death, both vegetable and animal life become part of the soil. During life, trees shed branches and twigs, due to storms, which all become part of the soil. Deciduous trees shed leaves and fruits or seeds annually. All vegetation has tremendous quantities of roots growing downward. These all finally disintegrate and become part of the soil in which they grew. All this growing tissue adds to and subtracts very little from the soil volume.

In addition to the soil increments from trees and from all other forms of plant life, prolific increments are added by animal life, in the form of dung while living and carcasses at death. All of this has become merged in the building-up processes of the earth's soil. "If you live to be 70 in the United States, your lifetime menu will have included 150 head of cattle, 26 sheep, 310 pigs, 225 lambs, 2,400 chickens, 26 acres of grains, and 50 acres of fruits and vegetables." (George Fuermann, in the *Houston Post*.)

This quotation calls attention to the vast volume of bodily excrements from human beings that go into the formation of topsoil. Man is, of course, a very small percentage of the total animal life which has flourished on this earth during past ages. The soils

Pre-Historic Forest Floors

change eventually to soil rocks through petrification and metamorphism—caused, at least partially, by agency of the mineral-laden, sub-surface waters percolating through it.

In the ocean, other forms of future rock strata are now being created by the animal life of corals and shellfish; they create limestones, marbles, and shales. Seaweeds and other vegetable growth also create future rock strata. All of these processes are going on today, just as they did in past ages.

These citations illustrate the fact that matter is created from radiant energy; but matter is also convertible into energy—such as by incandescence and, as recently demonstrated, by atomic-type bombs. The atom bomb has demolished older ideas, and appears to have established the equivalence of mass and energy. Two principles, the very cornerstones of the structure of modern science, hold that neither matter nor energy can be created or destroyed—but only altered in form.

The quantitative equivalence of mass and energy is based on the accepted—but unproven—suggestion by Albert Einstein that the relationship of energy, E, to mass, m, is shown in the equation $E = mc^2$, where c is the velocity of light.

If measured according to our conventional time-distance reckoning, the velocity of light is estimated at 186,300 miles per second. From this, there has come to be a general acceptance among men of science that mass is condensed energy—occluded or captured energy.

The atomic bomb provides a demonstration of the fact that when mass is suddenly converted back into energy, it disintegrates and apparently reverts to energy at about the speed of light. Originally, the materials were celestial energy rays. It is photosynthesis which brings about the condensation of energy rays into materials.

The equation for energy is:

$$E = \tfrac{1}{2} mv^2$$

where m is mass and v is velocity.

When mass is converted into energy, as in the atom bomb,

there is no longer mass, but only the energy of the liberated light rays, plus some unconverted energy that appears as fallout.

It is therefore suggested, quite logically, that the equivalence of energy and mass is more accurately expressed by the formula:

$$E = mc$$

Mathematical computation of the buildup of the earth's materials is now a possibility by substituting actual physical values for mass and energy in the formulas $E = mc$ or $m = \frac{E}{c}$. This will substantiate previous observations and deductions.

From celestial space we receive each day more than 2¼ billion B.T.U.'s (British Thermal Units), and from the sun approximately 14^{18} B.T.U's; an estimate of the energy arriving on the surface of the earth is given on page 178. In calculating the age of the earth (page 55) we have used a rate of buildup of one foot in 500 years (page 17), based on the Spur Ranch drilling; but at Ur of the Chaldeans and at Cnossus, Crete, as indicated above, the archeologists give figures for the earth buildup of one foot in about 125 years. They appear to have been areas of lush vegetation and much animal life.

These computations are based on "c"—the speed of light—being 186,300 miles per second. There are however known differences in the speeds of light rays. Red light rays have longer wave lengths and travel faster than violet light rays having shorter wave lengths.

"Altering the wave length of the light does change its velocity." (*The Encyclopaedia Britannica*, 1956, Vol. 14, page 620). Today, it is generally accepted in science that the light rays which become the atoms of lead travel at the same rate of speed as the light rays that become the atoms of hydrogen; but we know that their wave lengths are different and that, therefore, their speeds should be different.

Let us inquire into what becomes of the 14^{18} B.T.U's received daily on the earth from the sun. This energy must be stored up somewhere, because very little is rebroadcast back into space, and the temperature of the earth remains practically constant. One clue is the coal beds, which are huge layers of stored energy. The coal derives from vegetation, a fact which constitutes one more proof that vegetation is the result of the photosynthesis of the incoming radiant energy, the creator of all the materials of the earth.

Von Helmont's experiment, in the seventeenth century, is an interesting illustration of how the change from incoming radiant energy to earth materials occurs. He planted a 5-pound willow tree in 200 pounds of earth and added only water. At the end of five years the tree weighed 169 pounds, with practically no change in the weight of the earth materials in which it grew. It grew on air, water, and radiant energy from celestial space. Knowing that the increased weight of the tree, 164 pounds, represents energy according to the formula $E = mc$, we can compute, approximately, the amount of energy required to be added to the earth to create the materials of the tree. That tree is known to have absorbed CO_2 and to have given off hydrogen and oxygen; but what the relative amounts are is a question beyond the frontiers of present-day science.

Multiply that one tree by the countless billions which have grown on the earth's surface, and on dying, have returned materials to the earth—most of which they did not extract from it—and the nature of the building-up process of the earth's surface becomes apparent.

Trees, plants, and animals have been shown to eventually become parts of the soil rocks. Corals and sea shells become limestones. Parts of fishes make additions to bottom accumulations. But there is a preponderant rock—known as sandstone—that grows from sand. Sand appears to be created in the shallow seas directly from radiant energy.

Sand

LOOKING ABOUT us, we discover that sea sands occur throughout the world, in shallow ocean waters, around the coasts of all continents and islands—even coral islands. We discover that there are today tremendous areas of seacoast sands, which have not as yet turned into rocks, and which can be accounted for only by the theory of the creation of sand by the slow process of conversion of energy into matter.

We know that crystals of snowflakes are formed in the air (see page 223) by water, which is a mineral, so we may assume that by some similar, but not identical, metamorphosis the crystals of sand are formed in the shallow seas. They do not seem to form elsewhere.

Both snowflakes and sand crystals have this in common: they disintegrate at critical temperatures. Snowflakes change to water vapor above 32°F., and sand crystals change to gases at temperatures above the melting point of iron. Sandstone, having once been the sands created in the shallow seas, make up over 50% of the upper rock strata of the earth. The name "authogenetic sand" has been used by oceanographers to denote sands formed by direct precipitation from sea water through chemical reactions and in order to distinguish such sands from two other forms of sand of the seas, "terriginous" and "calcaranite," which signify, respectively, sands formed from disintegrating rocks and by shells and other fragments.

We can convert the sand back to light rays and can identify the wave lengths of the rays by the spectroscope; thus, it is not unreasonable to assume that what is now sand in our shallow seas and in the sandstone strata of our globe was once radiant energy in the form of light rays; and that the same general conditions of oceans, continents, islands, and sandy seacoast bottoms and beaches existed in former epochs when the now existing sandstone strata of our globe were being created.

Silica sands predominate in the shallow seas of the temperate zones, although considerable sections of ocean beaches consist mainly of broken and eroded sea shells.

Geological reports of the shallow seas and beaches of tropical islands and atolls show that the sands are mainly composed of granulated and pulverized materials derived from the mechanical and chemical disintegration of corals and sea shells, with very small amounts of silica sands.

There has never been any doubt that coral and shell formations have developed in the shallow seas from the growths of corals and shellfish. This material, which grew as part of animal life, consists of calcium carbonate and calcium-magnesium carbonate, and eventually becomes limestone and dolomite, respectively. Both are common rock strata of our earth.

III

The Creation of the Oceans

Four and a half billion years ago there were no earth strata and no waters of the earth. Now there are both. Their having been created is universally acknowledged, though the theories of creation have varied with time and place. It becomes perfectly reasonable to accept the new theory that the waters of the earth have been created, molecule by molecule, drop by drop, when we realize how other earth materials grow through photosynthesis from celestial radiant energy. The creation of water from radiant energy, transmitted and converted through vegetation and animals, and released by combustion, is in accordance with the laws of nature.

The sources from which the waters of the earth are created, are identified as (1) combustion of organic materials, (2) photosynthesis in vegetation, (3) chemical reactions in animals.

Combustion of Materials Creates Water

Having shown and verified that the land areas of the earth were created through photosynthesis and are growing in volume, we will now show with equally compelling evidence, that the waters of the earth are growing in volume and that they are being created in plants and animals; and that the water is mainly the product of the combustion of vegetation and animal materials.

In his book *Photosynthesis*, Eugene I. Rabinowitch states on page 1, "One trillion tons of organic carbon are produced each year." (Carbon fixation was a name previously used for photo-

The Creation of the Oceans 261

synthesis.) "The best estimate for the rate of the total carbon dioxide reduction on earth is a turnover of 10^{11} tons of carbon per year; more than ½ of which is contributed by the flora of the oceans." (*The Encyclopaedia Britannica*, 1959, Vol.17, page 848). Production and reduction are thus seen to be equal, 10^{11} and one trillion being different notations for the same number!

Vegetation shows constant growth and it all eventually disappears by decay, or rotting, which is slow combustion. As much dies each year, on the average, as is replaced by new vegetable growth; thus, one trillion tons of vegetable growth may be considered as being burned up by oxidation each year. This figure is one factor in the approximate rate of growth of the oceans. The quantity of vegetation that is converted into coal and lignite or that becomes petrified as stone is not essential for the following calculation.

In all plant materials the average ratio of the main chemical elements is indicated by the formula $C_6H_{12}O_6$ (see page 247, on minerals in vegetation). Approximately 7% of the weight of all vegetation is hydrogen, based on the atomic weights of the chemical elements. The combustion of each pound of hydrogen creates nine pounds of water vapor, as shown by the formula below. The vapor condenses as rain, and the waters flow through the rivers to the oceans.

Based on the above data we now have for consideration the yearly burning up of one trillion tons of vegetation, containing 7% hydrogen; 7% of one trillion tons is 70 billion tons, or (multiplied by 2,000) 140 trillion pounds of hydrogen; which, on combustion, creates nine times as much water, or 1,260 trillion pounds of water. Reduced to cubic feet, at 62.4 pounds of water per cubic feet, we have 20.2 trillion cubic feet of created water. There are 147½ billion cubic feet in a cubic mile; we thus have 137 cubic miles of water created each year by the combustion of vegetation. To visualize this amount of water, consider Lake Erie, which contains approximately 109 cubic miles, and Lake Superior with 2,927 cubic miles of water.

No less than 3,064,128,000 tons of water were created by the combustion of fossil fuels in the year 1958. It is roughly ½ of a cubic mile, and can be visualized as a lake with an average depth of twenty feet and covering 175 square miles. It is more water than is contained in the entire course of the Hudson River. This information is contained in a communication from the Bureau of Mines, U.S. Department of the Interior; it is based on *Worldwide Fuel Data 1958* and assumes the complete combustion of the fuels.

The equation representing the combustion of hydrogen:
$$2H_2 \text{ plus } O_2 \text{ equals } 2H_2O \text{ plus } 246{,}960 \text{ B.T.U.'s (heat)}$$
(4 lbs.) (32 lbs.) yields 36 lbs. of water vapor

The equation representing the combustion of carbon:
$$C \text{ plus } O_2 \text{ equals } CO_2 \text{ plus } 169{,}800 \text{ B.T.U.'s (heat)}$$
(12 lbs.) (32 lbs.) yields 44 lbs. of carbon dioxide gas

Based on a pound molecular weight of carbon, 12 lbs. C
hydrogen, 2 lbs. H_2
oxygen, 32 lbs. O_2

(From *American People's Encyclopedia*, 1954, Vol. 5, page 922)

When you burn a ten-gallon tankful of Esso regular gasoline in your automobile, you add 9.295 gallons of water to the oceans, according to a memorandum from Standard Oil (N.J.) Company's Research Department. Its weight is 61.51 lbs. and it contains 14% hydrogen.

A recent communication from the U.S. Coast and Geodetic Survey expresses the opinion that sea levels generally are rising at an average rate of 0.3 to 0.4 feet per century. Assuming this to be 0.0035 feet per year and with the area of all ocean seas being estimated at 139,495,122 square miles, then:

139,495,122 × 27,878,400 [sq. ft. per sq. mile] × 0.0035
=47 ¼ billion cubic feet, or about ⅕ of a cubic mile, being the extent to which the ocean level is rising.

This valued opinion indicates that the yearly growth rate of the oceans is practically stationary, while, as previously explained, ocean water increases at more than 137 cubic miles per year.

The Creation of the Oceans

This makes it logical to assume that the excess volume is being stored in the glacial ice of the polar regions.

The U.S. Coast and Geodetic Survey has estimated that the volume of ice now in the South Pole Ice Cap would make a layer 120 feet deep if spread uniformly around the earth. A depth of 120 feet of ice corresponds approximately to a layer of water 110 feet deep, and as the oceans occupy about 71% of the earth's surface, we have 110 divided by .71 = 155 feet. If all of the ice now accumulated in Antarctica's cold storage were to change back to water, the ocean levels would rise about 155 feet. Put in reverse order the ocean levels are about 155 feet lower than they would have been if there were no South Pole Ice Cap.

Sea levels have not remained constant. Determinations by Carbon 14 datings of materials taken from below sea level has shown a continuous rise in the ocean levels. The reports carried in scientific journals have generally attributed the rise in sea levels to the melting of the glaciers of a theoretical ice age, rather than this having been caused by the continuous creation of water.

We have records of both rising and receding sea levels. They appear to have fluctuated and to be lower now at the Bay of Fundy, Nova Scotia, and Long Island, New York, than they were in the past. The receding of the Red Sea caused the abandonment of a pre-Suez canal which was built about 2000 B.C., it was later extended but was discontinued in the eighth century A.D. (*Columbia Viking Desk Encyclopedia*, page 958).

The Production of Water

"DAVY IN 1817 discovered that when hydrogen and oxygen are passed through a tube, heated to temperature between 360° ånd 500° C. they combine to form water, without any violence and without light." (*The Encyclopaedia Britannica*, 1952, Vol. 6, page 98.)

Around 1780 Henry Cavendish, found that "When inflammable air and common air are exploded in a proper proportion, almost all the inflammable air and near one fifth of the common air . . . are turned into pure water." (From *Gaseous Combustion*

at High Pressures, by W.A. Bone and D.T.A. Townsend, 1929).
"Water may be produced by exploding a mixture of two volumes of hydrogen and one volume of oxygen at a temperature above 1190°F. . . . It may also be produced by passing hydrogen over the heated oxides of several of the metals, and in various other ways." (*The Encyclopaedia Britannica,* 1951, Vol. 29, page 13).

ANIMAL LIFE CREATES WATER. "All the hydrogen that unites with oxygen in our bodies forms water, and for the average person it amounts to about a pint a day." (*Water— Miracle of Nature,* by Thomson King, 1953.) The internal production of water seems to be the only way some animals get enough of it. The carpet moth is one of the bugs which drinks no water, yet lays eggs containing 80% water.

The camel as a creator of water is cited by John Eric Hill in *Natural History* magazine of October, 1946. He states: "The animal starch of glycogen, stored in the muscles, and the fat in the hump also provide water indirectly. When these are used by the body as energy, water of equivalent weight is produced. Thus, the fat of the hump—independent of the matter in the connective tissues—makes some eight gallons of water."

An excellent review of sources of water in vegetables and animals is given by S.M. Babcock in "Metabolitic Water—Its Production and Role in Vital Phenomena," in *Research Bulletin,* 22 (1912), Wisconsin Agricultural Experiment Station. Authors of biology and physiology books, when discussing metabolism, give chemical equations for the water and carbon dioxide given off in the oxidation of carbohydrates and fats in the body cells.

PHOTOSYNTHESIS CREATES WATER. The growth of the vegetation of the earth is now accepted in theory as being the result of photosynthesis. Trees and vegetation are classed as carbohydrates, *i.e.,* composed mainly of carbon, oxygen, and hydrogen, combined through photosynthesis. By extending this theory one will be led to suggest that the hydrogen and oxygen, which are products of photosynthesis, were created during the

The Creation of the Oceans

growth of vegetation, and that they united to form water, H_2O. The theory here postulated is that water is created by photosynthesis. Water is created through the agency of vegetation, and vegetation depends on water. Water is found in trees 300 feet above the ground. Leaf-suction and root-pressure theories fail, as long as they adhere to the laws of physics, to account for the bringing of this water up and out of the ground. Moisture in trees moves both upward and downward—also sideways in both directions in the branches. (See *Plant Physiology*, by Meyer and Anderson, 1952.)

Trees absorb carbon dioxide, CO_2, and give off oxygen, O. Animals absorb oxygen and give off carbon dioxide. Thus, vegetation and animal life have cooperated as they have developed on the earth. In the web of life one depends upon the other. The growth processes are postulated as originally having been started by photosynthesis of those incoming celestial radiations of energy whose wave lengths are the same as the wave lengths of the atoms of carbon, oxygen, and hydrogen.

Electric currents in vegetation have a share in the formation of water from two products of photosynthesis, oxygen and hydrogen. A uni-directional electric current carries oxygen toward the positive pole and hydrogen toward the negative pole of any electrolytic solution. For a 300-foot tree top there is normally a difference of around ten thousand volts in electrical voltage between the ambient air of the tree top and the ground, and during thunderstorms the difference in voltage of the air between tree top and ground may increase to around a million volts. (*Physics of the Air*, by W. J. Humphreys, pages 397-98).

Air voltages become so high at times, and the resulting passage of electric current between air and ground is so heavy and concentrated, that it becomes visible as a violet light or flame. St. Elmo's fire at the tips of the ships' masts is an example. High air voltages are today bled off to the ground by means of metal conductors in order to circumvent sudden discharges of large volumes of electricity by lightning flashes.

Tree tops are normally electrically positive and the ground negative, but this condition is often found to be reversed, especially during thunder storms. Electric currents will flow wherever and whenever a difference in electrical potential exists and a conducting medium connects them. Tree saps are such conducting mediums. The liquid elements between wood and bark of trees are conductors of electricity.

In an electrolytic solution some, but not all, of the molecules are broken down into ions by the electric currents. The ions keep resociating into molecules. (The Arrhenius theory. See *Plant Physiology*, by Meyer and Anderson, page 12.) The oxygen and hydrogen ions, which are carried by the electric currents, will, under proper conditions, unite to form water, H_2O. The molecules of water form into drops of water and little drops of water make the mighty oceans.

Moisture in the ground soil is necessary to complete the electric circuits which conduct the electricity between air and ground. Moisture is required for the roots to make adequate conducting contacts with the ground. Damp soil is a good electrical conductor; dry soil becomes a di-electric, or insulator. When the ground soil becomes dry, the electric currents are reduced in volume; and when the electric currents cease to flow, the creative processes of photosynthesis cease.

Trees and other vegetation have been generally assumed to get their main water supply from the ground; but Spanish moss and tropical orchids, called air plants or epiphytes, have been assumed to obtain their water from the atmosphere. The latter grow on other plants or on grounded supports, and normally have holdfasts instead of roots. Mistletoe, which grows in a manner similar to the epiphytes, is classed as a parasite. It grows, like other plants, from photosynthesis; and it appears rational to assume that the water contained in it is also a product of photosynthesis.

A well-established theory holds that sugars are products of photosynthesis. By analogy, the photosynthesis of water appears to be an equally reasonable theory. For example: sugar and molasses (produced from the consecutive yearly crops of sugar cane, and the maple sugar concentrated from the saps of sugar maple trees) do not come up out of the ground. Today, few

The Creation of the Oceans

educated people believe that these products of photosynthesis come out of the ground. We know that the ground does not contain sugar and molasses.

The wood in trees is composed of carbon, hydrogen, and oxygen. The carbon does not come up out of the ground. It is produced by photosynthesis, some of it being absorbed from the carbon dioxide of the air.

Land and water are postulated as having been contemporary and cooperative products of creation during the past 4¼ billion years of the earth's existence and growth. Soil rocks have been previously accounted for and it has been explained that they result from various vegetable, animal, and mineral growths. The continuous creation of the land has resulted in the development of continents and islands in the oceans.

There appears to be a Balance of Nature. If water is created too fast, it floods the land and thus decreases the area for the manufacture of water by allowing less space for vegetation and animals. The coastal shelves of many continents seem to indicate that the oceans have encroached on the land areas and have thus cut back water production. The Atlantic shelf of the east coast of the United States has that appearance.

The World Almanac, 1958, page 507, quotes the U.S. Navy Hydrographic Office as follows: "Surrounding most of the continents is a shallow zone of varying width which represents underwater continuation of continental land masses. These continental shelves are connected to oceanic basins by continental slopes, which are characterized by much greater angles of slope than either the continental shelves or the floors of the oceanic basins. The continental shelf and continental slope make up the continental terrace."

The change in the angle of slope in the continental terrace gives us a clue to a remote period of time when the oceans increased the rates or speeds of their growth; this, in turn, indicates a corresponding increase in the vegetable and animal life upon which the creation of water depends.

What was a coastal shelf in one epoch of time may have been either a land area or a sea bottom during the next epoch—and then a coastal shelf again in the following epoch; for it must be remembered that the temporary sea levels of continents for any single epoch of time are determined mainly by the latitudes at which the land areas happen to arrive after a normal careen of the globe.

More water than is contained in the combined volumes of Lakes Erie and Ontario is being evaporated each year from the oceans and deposited as snow, which becomes glacial ice on the Antarctic continent. A major percentage of this water is assumed to be returned to the oceans by the yearly flow-off of icebergs and ablation. Notwithstanding this yearly loss, and according to the evidence previously cited, the volume of water in the oceans and the amount of water frozen in the polar glaciers are both increasing.

IV

A Cosmic Cycle in the Eternity of Time

Our earth is composed of materials which are made up of atoms.
The atoms group together to form molecules. In modern physics all atoms and molecules are considered to be in continuous vibratory motion. When a material body is heated, it generally expands in volume because the atoms and the molecules increase the speeds of their motions: "In general, when heat is applied to substance, molecules vibrate faster and move farther apart; at constant pressure this causes expansion. When heat is withdrawn contraction occurs. If heat is added continuously, the velocity of the molecules becomes so great that change of state occurs." (*Columbia Viking Desk Encyclopedia*, page 646.)

The escape of atoms and molecules into the atmosphere, due to excessive velocities of vibration, is apparent in the boiling of water and in the burning of materials. The escape of atoms occurs in radium, carbon 14, and ionium.

Materials of the earth can exist in their present form only as long as they are kept at temperatures below those which cause them to become incandescent. A basic tenet of modern physics is that neither materials nor energy can be annihilated or destroyed; they can only be changed from one state to the other.

All of the 92 basic chemical elements of the earth have definite properties relative to light rays. They have the same wave lengths as the various light rays. Every material of the earth can be identified by its wave length, when it is heated to incandescence. The mark of identification is its wave length on the spectrograph charts—the wave-length scale of the light charts. The wave lengths show which light rays the materials correspond to, which they came from, and to which they return when overheated.

Atoms and light rays are, therefore, seen to be mutually convertible. Materials, made of atoms, and the radiant energy rays of light from celestial space are mutually convertible. Hence, mass and energy, planets and stars are mutually convertible.

Astronomers have reported, from time to time, the appearance of new stars in the heavens and the sudden disappearance of old established stars. Star births, called supernova, occur when a faint star suddenly becomes a brilliant star. (Another such star birth will occur when our earth planet suddenly blazes forth in a million times its present reflected light.)

Photosynthesis and combustion are postulated as the causes of the continuous creation of the materials of the earth. Photosynthesis is a process by means of which the incoming energy rays from celestial space are converted into earth materials. Combustion of materials creates water, as outlined in the section on "The Creation of the Oceans."

Tropical fauna and flora usually differ from those of the cold regions. This makes it evident that each material body requires a surrounding atmosphere, temperature, environment, and protection suitable for its formation.

The earth collects the radiated essence of sun and stars in its vegetable, animal and mineral growth, under specific conditions of temperature and environment. But raise the temperature of any of the earth's materials to incandescence or above, and it immediately metamorphoses and becomes flying molecules, or gases, with apparent release of some atoms. In the atomic-type bombs nearly all of the materials are apparently returned to energy, which becomes dispersed in space.

The continuous buildup of the materials of the earth, from its core to its present surface, suggests a cycle of time for the growth and development of the earth. It is built up in a series of stratifications, one superimposed on the other. The number of these layers is estimated to be about one million.

We can readily perceive that the top layer is building up and enlarging; and that topsoil is being created by vegetation and

A Cosmic Cycle in Time

other materials. Our knowledge of earth materials below the surface is confined to the top four miles of these earth layers, these being all that has been penetrated and sampled by drillings. It is readily found that the chemical elements in the successive layers generally differ from those immediately above and below at any one location.

The time cycle for the development of this layered accumulation of superimposed earth materials is about 4½ billion years, and there is no known reason why the bottom layers of the earth's stratifications should have been created in any manner substantially different from the top layer which is now being created.

We have learned how to reverse the slow process of the creation of certain materials of the earth and return them explosively back to energy. The atomic-type bombs, when detonated, do just this. Each of those sudden releases of atomic energy is equivalent to a tiny element of the sun's fiery heat.

To date, the atomic bomb is known to be made of only two of the earth's chemical elements. Whether the other ninety basic chemical elements are fissionable or not has not yet been demonstrated.

It seems quite reasonable to conclude that if fission is effected with the use of the chemical elements uranium and thorium, or their isotopes, some of the other elements will also prove to be fissionable or fusible and bring about thermonuclear reactions.

The hydrogen bomb is considered to be a self-sustaining thermonuclear reaction, similar to that of the sun. Therefore, should a self-supporting thermonuclear reaction become established by the materials of the earth it will change the earth into a sun, or star.

It is postulated that at the moment fission or fusion is started with some of the more common elements of the earth's materials, all of the earth's materials will blow up—like a hydrogen bomb—and will be changed back into energy; and at that time our earth will become a sun or star, just one more of the countless billions of self-sustaining thermonuclear reactions now observable in celestial space.

A few years ago the question of the possibility of the earth's blowing up from the fission of the uranium isotope U-235 was referred to the Atomic Energy Commission. U-235, which was used to make the first atomic bomb, is found to occur naturally in rock outcroppings in proportion of one U-235 to about 140 U-238 atoms. In a communication dated September 27, 1963, the Commission quotes the official report of Henry D. Smyth, *Atomic Energy for Military Purposes*, section 12:10: "No self-sustaining chain reaction could be produced in a block of pure uranium metal, no matter how large, because of parasitic capture of the neutrons by U-238." This conclusion has been borne out by various theoretical calculations and also by direct experiment. The Commission mentions the tiny amount of U-235 in the earth and states: "This trace of fissionable material is always too diluted with neutron-greedy atoms to permit a chain reaction from propagating."

The difference between the energy of a star and that of a planet is that in the star the energy is free and flaming, while in a planet it is captured, frozen, or occluded. The wave lengths of the essences of each are the same.

Our earth has a core, possibly of fused materials resembling our moon, from which it has grown upward, epoch by epoch, to the present very large globe. Sonic soundings by echo waves, which bounce back from the core, indicate a change in the kinds of materials about half-way down between the surface and the center of the earth. The materials of the core differ from those of the stratifications.

The forecast—based on what is already known about atomic reactions—is that the earth will eventually blow up by fission or fusion, and become a star. Through this sudden burst of matter into flaming energy, it is assumed an electrical condition will be created in the ether, and will move the moon a greater distance away from the new star; the earth will become the moon's sun, and the distance between them will be something like 93 million miles instead of approximately 250 thousand miles, as at present.

We can expect this to happen because all heavenly bodies are charged with "like" electricity. Like repels like. This is what

A Cosmic Cycle in Time

prevents collisions of the celestial spheres. The distances separating celestial bodies are postulated to depend on the intensities of their charges of "like" electricity.

The planet Moon is then expected to develop and grow, through epochs of time, exactly as our earth has grown through the ages, through the million epochs or so begun and terminated by the recurrent careenings of the globe.

Meantime, the sun will reach a condition through its profligate dissipation of its energy, of being depleted to such an extent that it can no longer maintain its heat- and light-producing thermonuclear reactions and it will therefore collapse into fused materials and become something like our moon of today. The moon is pockmarked with craters, apparently caused by the eruptions of internal gases from the molten interior while its viscous surface cooled, shrank, and solidified. Scientific opinion tends to regard our sun as a red star which at one time was much hotter and was white.

This indicates the probability of the existence of a super cosmic cycle in the eternity of time, involving sun, earth, moon, and possibly other celestial bodies.

For approximately 4½ billion years our earth has been bombarded by a continuous rain of celestial energy in the forms of photons, quanta, and cosmic rays. This tremendous, quite unimaginable quantity of incoming energy has been occluded, by photosynthesis, into atoms and makes up the molecules of all the earth's materials. A tiny amount has been rebroadcast back into space. The buildup of the earth by the conversion of energy into materials is a snail-like process, in which celestial energy is transmuted into materials. The reconversion of matter back into energy may be explosive, as in the atom bomb, or slow, as in incandescence, carbon 14, and radium. When we detonate an atom-type bomb we release an infinitesimal fraction of this accumulated energy.

These deductive analyses and discoveries of natural and measurable forces which govern earth phenomena, may be described as the New Cosmic Geophysics. It recognizes the incoming energy radiations from celestial space as the primary forces of nature. These forces, together with the light and heat radiations from the sun, govern earth phenomena.

Incoming radiant energy is the cause of gravitation, of the rotation of the earth, and of the growth of the earth by continuous creation. This celestial energy is indirectly the cause of the earth's careening, for it creates the centrifugal force which, acting on a rotating ice cap thrown off center by the wobble of the earth, causes the roll-arounds of the globe at irregular recurrent intervals of time.

Scholarly thought and opinion in the field of historical geology have undergone a gradual change during the past one hundred and fifty years. New evidence has been discovered and classified. The records left for us to read, fossils, etc.—in all of the earth strata penetrated by drillings—are now better understood. This geological evidence fits very well into the theories of the new cosmic geophysics, and it aids in confirming the recurrent careenings of the globe.

The theory of the evolution of biological forms postulates that the variations and mutations (which occurred from time to time and were better adapted and more suited to the environment in which they had to live) were the final survivors; similarly, the development of a philosophical geological science will depend on the discovery of better methods for evaluating the records left for us to read in the fossils and in the rocks.

Basic progress occurs like a mutation or a sport. By its very nature it must be the product of the nonconformist.

The Earth Is a Great Stone Book

Each single layer of earth
Tells a story that's all its own;
The sands of the ancient beaches
Have changed into strata of stone.
The growths of corals and sea shells
Made limestones, marbles, and shales;
While animals, trees, and vegetable growths
Made rocks which also tell their tales.
Each stratum of earth was created
During one of the epochs of time;
Remains of life growths are now embedded
In rocks that were once dirt and slime.
The limestone strata of Himalaya
Grew in ocean waters' shoal;
The glacial markings in tropical lands
Are the epoch's ice cap scroll.
Siberian mammoths were buried alive,
Interrupting their tropic stroll;
And a tropical land of the previous age
Now lies ice-embalmed at the Pole.
For the earth is a great stone book
With strata of stone for pages;
In which we'll find if we look
The living record of ancient ages.

H. A. B.

Index

Abbot, Charles G., 177
Abraham, 139
Adams, L. H., 230
"Adam's Wood" in Siberia, 31
"Adelie Land," winds in, 31
Adkins, W. S., 69
Agassiz, Louis, 44-8
Amazon valley glaciers, 48
Antarctica, bottom pressures, 100; cycle of growth, 111; height of ice, 109; precipitation in, 110; vicious circle of growth, 113; volume of ice, 109
Appalachian Mountains, drill log, 84
Arctic Ocean drift, 117
arctic shelf, once dry land, 71
arctic tundra, 19; former climate, food supply, 20
Arnold, Hugh M., 115
Atlantic mid-ridge, 69
Atlantis, 15
atoms and light rays, 270; binding force of atoms, 221

Babcock, S. M., 264
Ball, R. S., 161
balloon analogy, 131
barometric pressures, 189
Bayer, Ernest, 252
Bering Strait, current flow, 117
Bibby, Geoffrey T., 27
billiard ball analogy, 122
Bone, W. A., 264
Bonneville, Lake, 44
Borhyena, 22
Brush, Charles F., 175

Buddington, A. F., 229
Burbank, Luther, 215
Byrd, Richard E., 99

calendars, 134
Canada, height of, 114
careen, axis of, 141; eastern pivot point of, 142
Carver, 40
Caspian Sea Depression, 62
Caspian Sea Epoch, 59
cataclysm's cause, 146
catastrophists, 15
celestial spheres, held apart, 198
Chad, Lake, 58
charged spheres, 194-7
chemical elements and corresponding light rays, 253
chemical elements in vegetation, 248
Chilean Nitrate Company, 248
chlorophyl, 218-22
clay, 64, 66
Cleland, H. F., 29
Cnossus, Crete, 18
coal, 86
comets, 204
compass, 244
compass, inductor, 244
Confucius, 15
continents, heights of, 7
Coon, Carleton S., 140
copper-bearing tree, 250
cores from ocean bottoms, 70, 72
cosmic cycle in universe, 269
creation, continuous, 220
Cuvier, Georges, 15, 26, 40, 50, 65

Daniels, H. A., 235
Darwin, G. H., 176
Dead Sea, 97
DeGeer's time scale, 65
Descartes, 174
diamond "pipes", 106
dinosaurs, in Texas, 27
Donnelly, Ignatius, 15, 139
Doomsday globe, 5
Deucalion, 14, 51
dynamic electricity, 183
dynamic repulsion, 181

earth, after careen, 140; age of, 55; bulge of, 125; cause of rotation, 185, 187; electric currents, 238; heat of earth's currents, 239; reasons for careening of, 146; speed variance from moon, 201; strata and epochs, 55; tides, 209; wobble of, 68, 118, 190
earthquake equation, 126
Edison Effect, 180
Egypt, escapes Flood, 138, 139
Einstein, Albert, 255
elliptical orbit, 199
energy, equivalence to mass, 178; incoming measured, 178; metamorphoses to matter, 180
English Channel, creation of, 97
epochs of time, 55
Eskimos, 143
ether, medium for electric forces, 175
ether stream, 175

falling bodies, deflected east, 191
Flinch, Vernon C., 102
fish graveyards, 90
figs, petrified, 31
flag pole analogy, 130
Flood, the, 139; balance of land and water, 267; deluge mechanics, 118; survival of civilizations, 137

fly-wheel formula, 127
folded mountain, Appalachian, 103
forests, fossil floors of, 225
fossils, date age of strata, 25; fossil fruit, 32; fossil jelly fish, 28; fossil trees, 29, 32-6
Fram, the, 117
Frankfurter, G. B., 250
fruits, ancestry of, 142
Fuerman, George, 254
Fundy, Bay of, fossil trees, 32

Galileo, 169
gamma rays, stopped, 184
Gardner, Martin, 50
Gilber, G. K., 44
Gilbert, William, 246
glass, 182
globe in hoop analogy, 131
Goodspeed, G. E., 229
Graham, John W., 86
Grant, Frank F., 229
gravity, force of "G" not measured, 165; not an attraction, 174; research project, 209-13; unproved assumptions, 162
Gravity Research Foundation, 176
great west wind, 190
Gregory, J. W., 96
Gulf Stream, 188
Guyots, drowned islands, 105
gyroscope, 129, 144

Hague, Arnold, 227
Hall, James, 38
Hennepin, 40
Herodotus, 148
Hichins, C. S., 228
Hill, John Eric, 264
Hot Springs, 236
Hough, Jack, L., 70
Hudson Bay Basin, 47; ice cap, 61
Humphreys, W. J., 193, 240, 265
Hutton, James, 228
hydrogen, combustion of, 262

Index

ice, luminous blue, 99
Ice Age, 44; Antarctica during, 43, 45, 109; ice bowls, 99; ice caps, 111; migration of present ice cap, 143; ice front, 112; ice mountain, 99; ice volcanoes, 99
icebergs, 112
Inca civilization, 141

Jeanette, the, 117
Joggins, Nova Scotia, fossil trees, 33
Johnson, Charles F., 174
Jourdain, P. E. B., 160
Judi, Mt., and the Ark, 141

Kaemffert, Waldemar, 177
King, Thomson, 264
Klein, Dr. Richard M., 247
Knowlton, F. H., 228
Koran, the, confirms the Flood, 141
Kuhn, T. S., 174

land hemisphere, 94
Lambert, Walter D., Dr., 209
Laurentian Shield, 47, 67, 115
Le Sage, of Geneva, 173
Libby, Williard F., 62
lichens, create soil, 234
life ages, 43
light rays, 183; as electric currents, 184; stopped by water, 184
limestone, creation of, 235
Little America, 110
lobsters, 24
Long Island, a moraine, 52
Lorentz, Hendrick A., 173-4
Lyell, Sir Charles, 26, 37, 38

Malay civilization, 142
mammoths, 19, 27, 46; Bereskovka, 20; food of, 21; graveyards of, 28
mammoth tree, 31
mastadons, 46
Mawson, Sir Douglas, 112

Meyer and Anderson (Plant Physiology), 265-6
migrations of peoples, 136
Miller, Hugh, 89
Miller, J. S., 175
Millikan, Robert A., 184
minerals; aragonite, 234; cadmium, 252; carbon, combustion of, 262; carbon 14, dating, 61; chromium, 252; cobalt, 252; copper, origin of, 251; gold, 92; tin, 252
Mississippi River, 41; Mississippi River Commission, 41; delta, 42
molecular hatrack, 222
moon, influence of planets on, 203
Morton, C. V., 21
mountains, creation of, 101; heights of, 107

Nansen Fridtjof, 117
National Science Foundation, 111
Newberry, J. J., 91
Newton, Sir Isaac, 159
Niagara Falls, 37; retreat of, 38, 40
Noah, 14, 51
Nordenskiöld, 30, 34, 231
North Pole ice conditions, 117

oases, caused by electric heat, 101
oceans, 93; creation of, 259; deflection of currents of, 188; depths of, 107; floors of, 93
oil, 88
"Old Red" sandstones, 89
oxygen 18, in shells, 71

Panama Canal, bulges in, 28, petrified oysters found in, 28
paracutin volcano, 240
photosynthesis, 218
pith ball experiments, 198
planets and stars, convertible, 270
plants, nocturnal growth of, 254

Plato, 15
Platt, Rutherford, 234
population growth, 137
Priests of Memphis, 148

Rabinowitch, Eugene I., 260
Rastall, R. H., 231
Red Sea, 96
"reefs" gold ores, 93
rhinoceroses, Velui River, 21
Rideout, George M., 176
rifts, 96
Riksmuseum, algae of former epoch, 69
Riley, Charles M., 92
rivers, 35; erode west banks, 190
rocks, artificially made, 233; chemical composition of, 249; coastal extrusions, 99; embryonic, 233; flow at Panama Canal, 115; formation by heat cycles, 234; granite, 228; hardpan becomes rock, 217; igneous rock theory, 228; magnetic, 85; oldest, 53; orgneous, 228-9; faults repaired by nature, 240; growth of, 233; upthrusted coastal, 98
Rocky Mountain drill log, 83
Runcorn, S. K., 86

Safest areas, 150
Saks, V. N., 70
salt, 92; domes and pillars of, 107
sand, creation of, 257
sand crystals, creation of, 258
satellites' rotations, 198-9
Saturn's rings, analogy, 132
seals, 24; seals and boulders uplifted, 98
Senmut's Tomb, 148
Sheep Mountain, 104
silica, transported in solution, 228

Simplon Tunnel, heat in, 237
Smith, William, 68
Smyth, Henry D., 272
snowflakes, creation of, 223
"soilrocks," 228
Solon, 15
South Pole Height, 113
Spur Ranch, deep drilling, 73
St. Anthony's Falls, retreat of, 40
St. Davis Valley, 39
stars and planets, composition of, 270
static electricity, 191
Strait of Gibralter currents, 188
Sudan Basin, 59; ice cap, 61
Sumatra and Malay civilization, 142
sun, rays penetrate opaque bodies, 185
super-hurricanes, 132
Superior, Lake, analogy, 8; black shales from, 115
supernova, 272
Sverdrup, H. V., 208, 237
Sydney coal fields, 34

table land, Table Mountain, 51
television, and creation compared, 219
Thomas, Charles W., 72
Thomasson, William, 15
thunderstorms, movement of, 190
tides, 205; estuary tides, 206
time scales, 134
titanium, 251
Tolmachoff, I. P., 22
Townsend, D. T. A., 264
Travertine, 234

Uddin, J. A., 73
uniformitarians, 16
universe, expanding, 203
Ur of the Chaldees, 17

Index

Vega, The Voyage of the, 30
vegetables, ancestry of, 142
volcanoes, 236
Von Helmont's experiments, 257
Von Sterneck's experiments, 164

water, creation of, 263
water hemisphere, 94
Weddell Seal, 98

weight, ratio of permeabilities, 185
weightlessness, 186
Western Pivot Point Axis, 141
Willis, Bailey, 96
Wilson Cloud Chamber, 223
Winston, W. M., 69
Wood Hill, New Siberian Islands, 34; Wood underground, 64
Woodward, Robert S., 37
Worcester, P. G., 50

www.AdventuresUnlimitedPress.com

LOST CITIES OF ATLANTIS, ANCIENT EUROPE & THE MEDITERRANEAN
by David Hatcher Childress

Atlantis! The legendary lost continent comes under the close scrutiny of maverick archaeologist David Hatcher Childress in this sixth book in the internationally popular *Lost Cities* series. Childress takes the reader in search of sunken cities in the Mediterranean; across the Atlas Mountains in search of Atlantean ruins; to remote islands in search of megalithic ruins; to meet living legends and secret societies. From Ireland to Turkey, Morocco to Eastern Europe, and around the remote islands of the Mediterranean and Atlantic, Childress takes the reader on an astonishing quest for mankind's past. Ancient technology, cataclysms, megalithic construction, lost civilizations and devastating wars of the past are all explored in this book. Childress challenges the skeptics and proves that great civilizations not only existed in the past, but the modern world and its problems are reflections of the ancient world of Atlantis.
524 PAGES. 6X9 PAPERBACK. ILLUSTRATED. BIBLIOGRAPHY & INDEX. $16.95. CODE: MED

LOST CITIES OF ANCIENT LEMURIA & THE PACIFIC
by David Hatcher Childress

Was there once a continent in the Pacific? Called Lemuria or Pacifica by geologists, Mu or Pan by the mystics, there is now ample mythological, geological and archaeological evidence to "prove" that an advanced and ancient civilization once lived in the central Pacific. Maverick archaeologist and explorer David Hatcher Childress combs the Indian Ocean, Australia and the Pacific in search of the surprising truth about mankind's past. Contains photos of the underwater city on Pohnpei; explanations on how the statues were levitated around Easter Island in a clockwise vortex movement; tales of disappearing islands; Egyptians in Australia; and more.
379 PAGES. 6X9 PAPERBACK. $14.95. ILLUSTRATED. CODE: LEM

LOST CITIES & ANCIENT MYSTERIES OF SOUTH AMERICA
by David Hatcher Childress

Rogue adventurer and maverick archaeologist David Hatcher Childress takes the reader on unforgettable journeys deep into deadly jungles, high up on windswept mountains and across scorching deserts in search of lost civilizations and ancient mysteries. Travel with David and explore stone cities high in mountain forests and hear fantastic tales of Inca treasure, living dinosaurs, and a mysterious tunnel system. Whether he is hopping freight trains, searching for secret cities, or just dealing with the daily problems of food, money, and romance, the author keeps the reader spellbound. Includes both early and current maps, photos, and illustrations, and plenty of advice for the explorer planning his or her own journey of discovery.
381 PAGES. 6X9 PAPERBACK. ILLUSTRATED. FOOTNOTES. BIBLIOGRAPHY. INDEX. $16.95. CODE: SAM

LOST CITIES & ANCIENT MYSTERIES OF AFRICA & ARABIA
by David Hatcher Childress

Across ancient deserts, dusty plains and steaming jungles, maverick archaeologist David Childress continues his world-wide quest for lost cities and ancient mysteries. Join him as he discovers forbidden cities in the Empty Quarter of Arabia; "Atlantean" ruins in Egypt and the Kalahari desert; a mysterious, ancient empire in the Sahara; and more. This is the tale of an extraordinary life on the road: across war-torn countries, Childress searches for King Solomon's Mines, living dinosaurs, the Ark of the Covenant and the solutions to some of the fantastic mysteries of the past.
423 PAGES. 6x9 PAPERBACK. ILLUSTRATED. FOOTNOTES & BIBLIOGRAPHY. $14.95. CODE: AFA

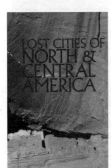

LOST CITIES OF NORTH & CENTRAL AMERICA
by David Hatcher Childress

Down the back roads from coast to coast, maverick archaeologist and adventurer David Hatcher Childress goes deep into unknown America. With this incredible book, you will search for lost Mayan cities and books of gold, discover an ancient canal system in Arizona, climb gigantic pyramids in the Midwest, explore megalithic monuments in New England, and join the astonishing quest for lost cities throughout North America. From the war-torn jungles of Guatemala, Nicaragua and Honduras to the deserts, mountains and fields of Mexico, Canada, and the U.S.A., Childress takes the reader in search of sunken ruins, Viking forts, strange tunnel systems, living dinosaurs, early Chinese explorers, and fantastic lost treasure. Packed with both early and current maps, photos and illustrations.
590 PAGES. 6x9 PAPERBACK. ILLUSTRATED. FOOTNOTES. BIBLIOGRAPHY. INDEX. $16.95. CODE: NCA

LOST CITIES OF CHINA, CENTRAL ASIA & INDIA
by David Hatcher Childress

Like a real life "Indiana Jones," maverick archaeologist David Childress takes the reader on an incredible adventure across some of the world's oldest and most remote countries in search of lost cities and ancient mysteries. Discover ancient cities in the Gobi Desert; hear fantastic tales of lost continents, vanished civilizations and secret societies bent on ruling the world; visit forgotten monasteries in forbidding snow-capped mountains with strange tunnels to mysterious subterranean cities! A unique combination of far-out exploration and practical travel advice, it will astound and delight the experienced traveler or the armchair voyager.
429 PAGES. 6x9 PAPERBACK. ILLUSTRATED. FOOTNOTES & BIBLIOGRAPHY. $14.95. CODE: CHI

TECHNOLOGY OF THE GODS
The Incredible Sciences of the Ancients
by David Hatcher Childress

Popular *Lost Cities* author David Hatcher Childress takes us into the amazing world of ancient technology, from computers in antiquity to the "flying machines of the gods." Childress looks at the technology that was allegedly used in Atlantis and the theory that the Great Pyramid of Egypt was originally a gigantic power station. He examines tales of ancient flight and the technology that it involved; how the ancients used electricity; megalithic building techniques; the use of crystal lenses and the fire from the gods; evidence of various high tech weapons in the past, including atomic weapons; ancient metallurgy and heavy machinery; the role of modern inventors such as Nikola Tesla in bringing ancient technology back into modern use; impossible artifacts; and more.
356 PAGES. 6x9 PAPERBACK. ILLUSTRATED. BIBLIOGRAPHY. $16.95. CODE: TGOD

ARK OF GOD
The Incredible Power of the Ark of the Covenant
By David Hatcher Childress

Childress takes us on an incredible journey in search of the truth about (and science behind) the fantastic biblical artifact known as the Ark of the Covenant. This object made by Moses at Mount Sinai—part wooden-metal box and part golden statue—had the power to create "lightning" to kill people, and also to fly and lead people through the wilderness. The Ark of the Covenant suddenly disappears from the Bible record and what happened to it is not mentioned. Was it hidden in the underground passages of King Solomon's temple and later discovered by the Knights Templar? Was it taken through Egypt to Ethiopia as many Coptic Christians believe? Childress looks into hidden history, astonishing ancient technology, and a 3,000-year-old mystery that continues to fascinate millions of people today. Color section.
420 Pages. 6x9 Paperback. Illustrated. $22.00 Code: AOG

LOST CONTINENTS & THE HOLLOW EARTH
I Remember Lemuria and the Shaver Mystery
by David Hatcher Childress & Richard Shaver

A thorough examination of the early hollow earth stories of Richard Shaver and the fascination that fringe fantasy subjects such as lost continents and the hollow earth have had for the American public. Shaver's rare 1948 book *I Remember Lemuria* is reprinted in its entirety, and the book is packed with illustrations from Ray Palmer's *Amazing Stories* magazine of the 1940s. Palmer and Shaver told of tunnels running through the earth—tunnels inhabited by the Deros and Teros, humanoids from an ancient spacefaring race that had inhabited the earth, eventually going underground, hundreds of thousands of years ago. Childress discusses the famous hollow earth books and delves deep into whatever reality may be behind the stories of tunnels in the earth. Operation High Jump to Antarctica in 1947 and Admiral Byrd's bizarre statements, tunnel systems in South America and Tibet, the underground world of Agartha, the belief of UFOs coming from the South Pole, more.
344 PAGES. 6x9 PAPERBACK. $16.95. CODE: LCHE

YETIS, SASQUATCH & HAIRY GIANTS
By David Hatcher Childress
Childress takes the reader on a fantastic journey across the Himalayas to Europe and North America in his quest for Yeti, Sasquatch and Hairy Giants. Childress begins with a discussion of giants and then tells of his own decades-long quest for the Yeti in Nepal, Sikkim, Bhutan and other areas of the Himalayas, and then proceeds to his research into Bigfoot, Sasquatch and Skunk Apes in North America. Chapters include: The Giants of Yore; Giants Among Us; Wildmen and Hairy Giants; The Call of the Yeti; Kanchenjunga Demons; The Yeti of Tibet, Mongolia & Russia; Bigfoot & the Grassman; Sasquatch Rules the Forest; Modern Sasquatch Accounts; more. Includes a 16-page color photo insert of astonishing photos!
360 pages. 5x9 Paperback. Illustrated. Bibliography. Index. $18.95. Code: YSHG

SECRETS OF THE HOLY LANCE
The Spear of Destiny in History & Legend
by Jerry E. Smith
Secrets of the Holy Lance traces the Spear from its possession by Constantine, Rome's first Christian Caesar, to Charlemagne's claim that with it he ruled the Holy Roman Empire by Divine Right, and on through two thousand years of kings and emperors, until it came within Hitler's grasp—and beyond! Did it rest for a while in Antarctic ice? Is it now hidden in Europe, awaiting the next person to claim its awesome power? Neither debunking nor worshiping, *Secrets of the Holy Lance* seeks to pierce the veil of myth and mystery around the Spear.
312 PAGES. 6x9 PAPERBACK. ILLUSTRATED. $16.95. CODE: SOHL

THE CRYSTAL SKULLS
Astonishing Portals to Man's Past
by David Hatcher Childress and Stephen S. Mehler
Childress introduces the technology and lore of crystals, and then plunges into the turbulent times of the Mexican Revolution form the backdrop for the rollicking adventures of Ambrose Bierce, the renowned journalist who went missing in the jungles in 1913, and F.A. Mitchell-Hedges, the notorious adventurer who emerged from the jungles with the most famous of the crystal skulls. Mehler shares his extensive knowledge of and experience with crystal skulls. Having been involved in the field since the 1980s, he has personally examined many of the most influential skulls, and has worked with the leaders in crystal skull research. Color section.
294 pages. 6x9 Paperback. Illustrated. $18.95. Code: CRSK

THE LAND OF OSIRIS
An Introduction to Khemitology
by Stephen S. Mehler
Was there an advanced prehistoric civilization in ancient Egypt? Were they the people who built the great pyramids and carved the Great Sphinx? Did the pyramids serve as energy devices and not as tombs for kings? Chapters include: Egyptology and Its Paradigms; Khemitology—New Paradigms; Asgat Nefer—The Harmony of Water; Khemit and the Myth of Atlantis; The Extraterrestrial Question; more.
272 PAGES. 6X9 PAPERBACK. ILLUSTRATED. COLOR SECTION. BIBLIOGRAPHY. $18.95. CODE: LOOS

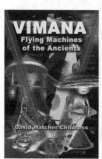

VIMANA:
Flying Machines of the Ancients
by David Hatcher Childress
According to early Sanskrit texts the ancients had several types of airships called vimanas. Like aircraft of today, vimanas were used to fly through the air from city to city; to conduct aerial surveys of uncharted lands; and as delivery vehicles for awesome weapons. David Hatcher Childress, popular *Lost Cities* author and star of the History Channel's long-running show Ancient Aliens, takes us on an astounding investigation into tales of ancient flying machines. In his new book, packed with photos and diagrams, he consults ancient texts and modern stories and presents astonishing evidence that aircraft, similar to the ones we use today, were used thousands of years ago in India, Sumeria, China and other countries. Includes a 24-page color section.

408 Pages. 6x9 Paperback. Illustrated. $22.95. Code: VMA

MIND CONTROL AND UFOS:
Casebook on Alternative 3
By Jim Keith
Keith's classic investigation of the Alternative 3 scenario as it first appeared on British television over 20 years ago. Keith delves into the bizarre story of Alternative 3, including mind control programs, underground bases not only on the Earth but also on the Moon and Mars, the real origin of the UFO problem, the mysterious deaths of Marconi Electronics employees in Britain during the 1980s, the Russian-American superpower arms race of the 50s, 60s and 70s as a massive hoax, more.

248 Pages. 6x9 Paperback. Illustrated. $14.95. Code: MCUF

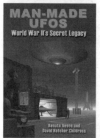

MAN-MADE UFOS
WWII's Secret Legacy
By Renato Vesco & David Hatcher Childress
The classic book on suppressed technology: the early "flying saucer" technology of Nazi Germany and the genesis of man-made UFOs. Examined in detail are secret underground airfields and factories; German secret weapons; "suction" aircraft; the origin of NASA; gyroscopic stabilizers and engines; the secret Marconi aircraft factory in South America; and more. Chapters include: June 24, 1947: A Day to Remember; The True Story of Project Blue Book; Mysterious Night Lights Over the Rhineland; Revolutionary German Anti-Aircraft Weaponry; How To Do the Impossible; Marconi's Secret Saucer Base; more.

524 Pages. 6x9 Paperback. Illustrated. $22.95. Code: MMU

ROSWELL AND THE REICH
By Joseph P. Farrell
Farrell comes to a radically different scenario of what happened in Roswell in July 1947, and why the US military has continued to cover it up to this day. Farrell presents a fascinating case that what crashed may have been representative of an independent postwar Nazi power—an extraterritorial Reich monitoring its old enemy, America, and the continuing development of the very technologies confiscated from Germany at the end of the War.

540 pages. 6x9 Paperback. $19.95. Code: RWR

IN SECRET MONGOLIA
by Henning Haslund
Haslund takes us into the barely known world of Mongolia of 1921, a land of god-kings, bandits, vast mountain wilderness and a Russian army running amok. Starting in Peking, Haslund journeys to Mongolia as part of a mission to establish a Danish butter farm in a remote corner of northern Mongolia. With Haslund we meet the "Mad Baron" Ungern-Sternberg and his renegade Russian army, the many characters of Urga's fledgling foreign community, and the last god-king of Mongolia, Seng Chen Gegen, the fifth reincarnation of the Tiger god and the "ruler of all Torguts." Aside from the esoteric and mystical material, there is plenty of just plain adventure: Haslund encounters a Mongolian werewolf; is ambushed along the trail; escapes from prison and fights terrifying blizzards; more.
374 PAGES. 6x9 PAPERBACK. ILLUSTRATED. BIBLIOGRAPHY & INDEX. $16.95. CODE: ISM

MEN & GODS IN MONGOLIA
by Henning Haslund
Haslund takes us to the lost city of Karakota in the Gobi desert. We meet the Bodgo Gegen, a god-king in Mongolia similar to the Dalai Lama of Tibet. We meet Dambin Jansang, the dreaded warlord of the "Black Gobi." Haslund and companions journey across the Gobi desert by camel caravan; are kidnapped and held for ransom; witness initiation into Shamanic societies; meet reincarnated warlords; and experience the violent birth of "modern" Mongolia.
358 PAGES. 6x9 PAPERBACK. ILLUSTRATED. INDEX. $18.95. CODE: MGM

MYSTERIES OF ANCIENT SOUTH AMERICA
by Harold T. Wilkins
Wilkins digs into old manuscripts and books to bring us some truly amazing stories of South America: a bizarre subterranean tunnel system; lost cities in the remote border jungles of Brazil; cataclysmic changes that shaped South America; and other strange stories from one of the world's great researchers. Chapters include: Dead Cities of Ancient Brazil, The Jungle Light that Shines by Itself, The Missionary Men in Black: Forerunners of the Great Catastrophe, The Sign of the Sun: The World's Oldest Alphabet, The Atlanean "Subterraneans" of the Incas, Tiahuanacu and the Giants, more.
236 PAGES. 6x9 PAPERBACK. ILLUSTRATED. INDEX. $14.95. CODE: MASA

SECRET CITIES OF OLD SOUTH AMERICA
by Harold T. Wilkins
The reprint of Wilkins' classic book, first published in 1952, claiming that South America was Atlantis. Chapters include Mysteries of a Lost World; Atlantis Unveiled; Red Riddles on the Rocks; South America's Amazons Existed!; The Mystery of El Dorado and Gran Payatiti—the Final Refuge of the Incas; Monstrous Beasts of the Unexplored Swamps & Wilds; Weird Denizens of Antediluvian Forests; New Light on Atlantis from the World's Oldest Book; The Mystery of Old Man Noah and the Arks; and more.
438 PAGES. 6x9 PAPERBACK. ILLUSTRATED. BIBLIOGRAPHY & INDEX. $16.95. CODE: SCOS

www.AdventuresUnlimitedPress.com

LOST CITIES & ANCIENT MYSTERIES OF THE SOUTHWEST
By David Hatcher Childress

Join David as he starts in northern Mexico and then to west Texas amd into New Mexico where he stumbles upon a hollow mountain with a billion dollars of gold bars hidden deep inside it! In Arizona he investigates tales of Egyptian catacombs in the Grand Canyon, cruises along the Devil's Highway, and tackles the century-old mystery of the Lost Dutchman mine. In California Childress checks out the rumors of mummified giants and weird tunnels in Death Valley—It's a full-tilt blast down the back roads of the Southwest in search of the weird and wondrous mysteries of the past!
486 Pages. 6x9 Paperback. Illustrated. $19.95. Code: LCSW

AXIS OF THE WORLD
The Search for the Oldest American Civilization
by Igor Witkowski

Witkowski's research reveals remnants of a high civilization that was able to exert its influence on almost the entire planet, and did so with full consciousness. Sites around South America show that this was a place where they built their crowning achievements. Easter Island, in the southeastern Pacific, constitutes one of them. The Rongo-Rongo language that developed there points westward to the Indus Valley. Taken together, the facts presented provide new proof that an antediluvian civilization flourished several millennia ago.
220 pages. 6x9 Paperback. Illustrated. $18.95. Code: AXOW

SECRETS OF THE MYSTERIOUS VALLEY
by Christopher O'Brien

No other region in North America features the variety and intensity of unusual phenomena found in the world's largest alpine valley, the San Luis Valley of Colorado and New Mexico. Since 1989, O'Brien has documented thousands of high-strange accounts that report UFOs, ghosts, crypto-creatures, cattle mutilations, and more, along with portal areas, secret underground bases and covert military activity. Hundreds of animals have been found strangely slain during waves of anomalous aerial craft sightings. Is the government directly involved? Are there underground bases here?
460 PAGES. 6x9 PAPERBACK. ILLUSTRATED. BIBLIOGRAPHY. $19.95. CODE: SOMV

PIRATES & THE LOST TEMPLAR FLEET
by David Hatcher Childress

The lost Templar fleet was originally based at La Rochelle in southern France, but fled to the deep fiords of Scotland upon the dissolution of the Order by King Phillip. This banned fleet of ships was later commanded by the St. Clair family of Rosslyn Chapel. St. Clair and his Templars made a voyage to Canada in the year 1398 AD, nearly 100 years before Columbus! Chapters include: 10,000 Years of Seafaring; The Templars and the Assassins; The Lost Templar Fleet and the Jolly Roger; Maps of the Ancient Sea Kings; Pirates, Templars and the New World; Christopher Columbus—Secret Templar Pirate?; Later Day Pirates and the War with the Vatican; Pirate Utopias and the New Jerusalem; more.
320 PAGES. 6x9 PAPERBACK. ILLUSTRATED. BIBLIOGRAPHY. $16.95. CODE: PLTF

EYE OF THE PHOENIX
Mysterious Visions and Secrets of the American Southwest
by Gary David

GaryDavid explores enigmas and anomalies in the vast American Southwest. Contents includes: The Great Pyramids of Arizona; Meteor Crater—Arizona's First Bonanza?; Chaco Canyon—Ancient City of the Dog Star; Phoenix—Masonic Metropolis in the Valley of the Sun; Along the 33rd Parallel—A Global Mystery Circle; The Flying Shields of the Hopi Katsinam; Is the Starchild a Hopi God?; The Ant People of Orion—Ancient Star Beings of the Hopi; Serpent Knights of the Round Temple; The Nagas—Origin of the Hopi Snake Clan?; The Tau (or T-shaped) Cross—Hopi/Maya/Egyptian Connections; The Hopi Stone Tablets of Techqua Ikachi; The Four Arms of Destiny—Swastikas in the Hopi World of the End Times; and more.
348 pages. 6x9 Paperback. Illustrated. $16.95. Code: EOPX

THE ORION ZONE
Ancient Star Cities of the American Southwest
by Gary A. David

This book on ancient star lore explores the mysterious location of Pueblos in the American Southwest, circa 1100 AD, that appear to be a mirror image of the major stars of the Orion constellation. Chapters include: Leaving Many Footprints—The Emergence and Migrations of the Anazazi; The Sky Over the Hopi Villages; Orion Rising in the Dark Crystal; The Cosmo-Magical Cities of the Anazazi; Windows Onto the Cosmos; To Calibrate the March of Time; They Came from Across the Ocean—The Patki (Water) Clan and the Snake Clan of the Hopi; Ancient and Mysterious Monuments; Beyond That Fiery Day; more.
346 pages. 6x9 Paperback. Illustrated. $19.95. Code: OZON

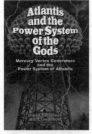

ATLANTIS & THE POWER SYSTEM OF THE GODS
by David Hatcher Childress and Bill Clendenon

Childress' fascinating analysis of Nikola Tesla's broadcast system in light of Edgar Cayce's "Terrible Crystal" and the obelisks of ancient Egypt and Ethiopia. Includes: Atlantis and its crystal power towers that broadcast energy; how these incredible power stations may still exist today; inventor Nikola Tesla's nearly identical system of power transmission; Mercury Proton Gyros and mercury vortex propulsion; more. Richly illustrated, and packed with evidence that Atlantis not only existed—it had a world-wide energy system more sophisticated than ours today.
246 PAGES. 6x9 PAPERBACK. ILLUSTRATED. $15.95. CODE: APSG

PATH OF THE POLE
by Charles Hapgood

Hapgood researched Antarctica, ancient maps and the geological record to conclude that the Earth's crust has slipped in the inner core many times in the past, changing the position of the pole. *Path of the Pole* discusses the various "pole shifts" in Earth's past, giving evidence for each one, and moves on to possible future pole shifts. Packed with illustrations, this is the sourcebook for many other books on cataclysms and pole shifts.
356 PAGES. 6x9 PAPERBACK. ILLUSTRATED. $16.95. CODE: POP

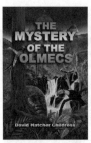

THE MYSTERY OF THE OLMECS
by David Hatcher Childress
The Olmecs were not acknowledged to have existed as a civilization until an international archeological meeting in Mexico City in 1942. Now, the Olmecs are slowly being recognized as the Mother Culture of Mesoamerica, having invented writing, the ball game and the "Mayan" Calendar. But who were the Olmecs? Where did they come from? What happened to them? How sophisticated was their culture? Why are many Olmec statues and figurines seemingly of foreign peoples such as Africans, Europeans and Chinese? Is there a link with Atlantis? In this heavily illustrated book, join Childress in search of the lost cities of the Olmecs! Chapters include: The Mystery of Quizuo; The Mystery of Transoceanic Trade; The Mystery of Cranial Deformation; more.
296 PAGES. 6x9 PAPERBACK. COLOR SECTION. $20.00. CODE: MOLM

SUNKEN REALMS
A Survey of Underwater Ruins Around the World
By Karen Mutton
Australian researcher Mutton starts with the underwater cities in the Mediterranean, and then moves into Europe and the Atlantic. She continues with chapters on the Caribbean and then moves through the extensive sites in the Pacific and Indian Oceans. Places covered in this book include: Tartessos; Cadiz; Morocco; Alexandria; Cyprus; Malta; Thule & Hyperborea; Celtic Realms Lyonesse, Ys, and Hy Brasil; Canary and Azore Islands; Bahamas; Cuba; Bermuda; Mexico; Peru; Micronesia; California; Japan; Indian Ocean; Sri Lanka Land Bridge; India; Sumer; Lake Titicaca; more.
320 Pages. 6x9 Paperback. Illustrated. $20.00. Code: SRLM

ATLANTIS IN SPAIN
A Study of the Ancient Sun Kingdoms of Spain
by E.M. Whishaw
First published in 1928, this classic book is a study of the megaliths of Spain, ancient writing, cyclopean walls, sun worshipping empires, hydraulic engineering, and sunken cities. An extremely rare book, it was out of print for 60 years. Learn about the Biblical Tartessus; an Atlantean city at Niebla; the Temple of Hercules and the Sun Temple of Seville; Libyans and the Copper Age; more. Profusely illustrated with photos, maps and drawings.
284 PAGES. 6x9 PAPERBACK. ILLUSTRATED. $15.95. CODE: AIS

RIDDLE OF THE PACIFIC
by John Macmillan Brown
Oxford scholar Brown's classic work on lost civilizations of the Pacific is now back in print! John Macmillan Brown was an historian and New Zealand's premier scientist when he wrote about the origins of the Maoris. After many years of travel thoughout the Pacific studying the people and customs of the south seas islands, he wrote *Riddle of the Pacific* in 1924. The book is packed with rare turn-of-the-century illustrations. Don't miss Brown's classic study of Easter Island, ancient scripts, megalithic roads and cities, more. Brown was an early believer in a lost continent in the Pacific.
460 PAGES. 6x9 PAPERBACK. ILLUSTRATED. $16.95. CODE: SOA

www.AdventuresUnlimitedPress.com

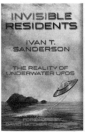

INVISIBLE RESIDENTS
The Reality of Underwater UFOS
by Ivan T. Sanderson

Sanderson puts forward the curious theory that "OINTS"—Other Intelligences—live under the Earth's oceans. This underwater, parallel, civilization may be twice as old as Homo sapiens, he proposes, and may have "developed what we call space flight." Sanderson postulates that the OINTS are behind many UFO sightings as well as the mysterious disappearances of aircraft and ships in the Bermuda Triangle. What better place to have an impenetrable base than deep within the oceans of the planet? Sanderson offers here an exhaustive study of USOs (Unidentified Submarine Objects) observed in nearly every part of the world.

298 PAGES. 6x9 PAPERBACK. ILLUSTRATED. $16.95. CODE: INVS

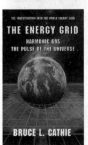

THE ENERGY GRID
Harmonic 695, The Pulse of the Universe
by Captain Bruce Cathie

This is the breakthrough book that explores the incredible potential of the Energy Grid and the Earth's Unified Field all around us. Bruce Cathie has been the premier investigator into the amazing potential of the infinite energy that surrounds our planet every microsecond. Cathie investigates the Harmonics of Light and how the Energy Grid is created. In this amazing book are chapters on UFO Propulsion, Nikola Tesla, Unified Equations, the Mysterious Aerials, Pythagoras & the Grid, Nuclear Detonation and the Grid, Maps of the Ancients, an Australian Stonehenge examined, more.

255 PAGES. 6x9 TRADEPAPER. ILLUSTRATED. $15.95. CODE: TEG

THE BRIDGE TO INFINITY
Harmonic 371244
by Captain Bruce Cathie

Cathie has popularized the concept that the earth is crisscrossed by an electromagnetic grid system that can be used for anti-gravity, free energy, levitation and more. The book includes a new analysis of the harmonic nature of reality, acoustic levitation, pyramid power, harmonic receiver towers and UFO propulsion. It concludes that today's scientists have at their command a fantastic store of knowledge with which to advance the welfare of the human race.

204 PAGES. 6x9 TRADEPAPER. ILLUSTRATED. $14.95. CODE: BTF

THE HARMONIC CONQUEST OF SPACE
by Captain Bruce Cathie

Chapters include: Mathematics of the World Grid; the Harmonics of Hiroshima and Nagasaki; Harmonic Transmission and Receiving; the Link Between Human Brain Waves; the Cavity Resonance between the Earth; the Ionosphere and Gravity; Edgar Cayce—the Harmonics of the Subconscious; Stonehenge; the Harmonics of the Moon; the Pyramids of Mars; Nikola Tesla's Electric Car; the Robert Adams Pulsed Electric Motor Generator; Harmonic Clues to the Unified Field; and more. Also included are tables showing the harmonic relations between the earth's magnetic field, the speed of light, and anti-gravity/gravity acceleration at different points on the earth's surface.

248 PAGES. 6x9. PAPERBACK. ILLUSTRATED. $16.95. CODE: HCS

ANCIENT TECHNOLOGY IN PERU & BOLIVIA
By David Hatcher Childress
Childress speculates on the existence of a sunken city in Lake Titicaca and reveals new evidence that the Sumerians may have arrived in South America 4,000 years ago. He demonstrates that the use of "keystone cuts" with metal clamps poured into them to secure megalithic construction was an advanced technology used all over the world, from the Andes to Egypt, Greece and Southeast Asia. He maintains that only power tools could have made the intricate articulation and drill holes found in extremely hard granite and basalt blocks in Bolivia and Peru, and that the megalith builders had to have had advanced methods for moving and stacking gigantic blocks of stone, some weighing over 100 tons.
340 Pages. 6x9 Paperback. Illustrated.. $19.95 Code: ATP

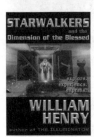

STARWALKERS AND THE DIMENSION OF THE BLESSED
By William Henry
Explore: The Egyptian belief that interdimensional beings of light created humanity; The meaning behind the 'reed,' the key term constantly repeated over thousands of years in these global myths; How the Place of Reeds, is an allegory for the opening of a gate to another realm; How the human body is capable of producing a spiritual substance that is made of space-time and through which one can see other times and places; more. Chapters include: A Cosmic Species; The Dimension of the Blessed; The Field; Up Out of Egypt; The Blessed Falcons; The Divine Spark of the Blessed; Atlantis: The Blessed Land; The Sea at the End of the World; Manna and the Blessed Realm; Blessed Sirius; Gilgamesh & Sirius; more.
270 Pages. 6x9 Paperback. Illustrated. $16.95. Code: SDOB

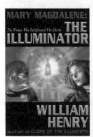

MARY MAGDALENE: THE ILLUMINATOR
The Woman Who Enlightened the Christ
By William Henry
Henry explores the core of the mysteries of Mary Magdalene to study knowledge of the 'ultimate secret' of the Tower or Ladder to God, also called the Stairway to Heaven; The alchemical secrets of Mary Magdalene's anointing oil and how it transformed Jesus; The Magdalene's connection to Ishtar, Isis and other ancient goddesses; The reality of an extraterrestrial presence in the Bible and Gnostic Christian texts; How the Knights Templar encoded the secret teaching of Jesus and Mary Magdalene in religious graffiti at Domme, France; more.
304 pages. 6x9 Paperback. Illustrated. $16.95. Code: MMTI

ORACLE OF THE ILLUMINATI
Coincidence, Cocreation, Contact
By William Henry
Investigative mythologist Henry on the secret codes, oracles and technology of the ancient Illuminati. His primary mission is finding and interpreting ancient gateway stories which feature advanced technology for raising spiritual vibration and increasing our body's innate healing ability. Chapters include: From Cloak to Oracle; The Return of Sophia; The Cosmic G-Spot Stimulator; The Hymn of the Pearl; The Realm of the Illuminati; Francis Bacon: Oracle; Abydos and the Head of Sophia; Enki and the Flower of Light; The God Head and the Dodecahedron; The Star Walker; The Big Secret; more.
243 Pages. 6x9 Paperback. Illustrated. $16.95. Code: ORIL

HIDDEN FINANCE, ROGUE NETWORKS & SECRET SORCERY
The Fascist International, 9/11, & Penetrated Operations
By Joseph P. Farrell

Pursuing his investigations of high financial fraud, international banking, hidden systems of finance, black budgets and breakaway civilizations, Farrell investigates the theory that there were not *two* levels to the 9/11 event, but *three*. He says that the twin towers were downed by the force of an exotic energy weapon, one similar to the Tesla energy weapon suggested by Dr. Judy Wood, and ties together the tangled web of missing money, secret technology and involvement of portions of the Saudi royal family. Farrell unravels the many layers behind the 9-11 attack, layers that include the Deutschebank, the Bush family, the German industrialist Carl Duisberg, Saudi Arabian princes and the energy weapons developed by Tesla before WWII.
296 Pages. 6x9 Paperback. Illustrated. $19.95. Code: HFRN

THRICE GREAT HERMETICA AND THE JANUS AGE
By Joseph P. Farrell

What do the Fourth Crusade, the exploration of the New World, secret excavations of the Holy Land, and the pontificate of Innocent the Third all have in common? Answer: Venice and the Templars. What do they have in common with Jesus, Gottfried Leibniz, Sir Isaac Newton, Rene Descartes, and the Earl of Oxford? Answer: Egypt and a body of doctrine known as Hermeticism. The hidden role of Venice and Hermeticism reached far and wide, into the plays of Shakespeare (a.k.a. Edward DeVere, Earl of Oxford), into the quest of the three great mathematicians of the Early Enlightenment for a lost form of analysis, and back into the end of the classical era, to little known Egyptian influences at work during the time of Jesus.
354 Pages. 6x9 Paperback. Illustrated. $19.95. Code: TGHJ

THE THIRD WAY
The Nazi International, European Union, & Corporate Fascism
By Joseph P. Farrell

Pursuing his investigations of high financial fraud, international banking, hidden systems of finance, black budgets and breakaway civilizations, Farrell continues his examination of the post-war Nazi International, an "extra-territorial state" without borders or capitals, a network of terrorists, drug runners, and people in the very heights of financial power willing to commit financial fraud in amounts totaling trillions of dollars. Breakaway civilizations, black budgets, secret technology, occult rituals, international terrorism, giant corporate cartels, patent law and the hijacking of nature: Farrell explores 'the business model' of the post-war Axis elite.
364 Pages. 6x9 Paperback. Illustrated. $19.95. Code: TTW

COVERT WARS AND BREAKAWAY CIVILIZATIONS
By Joseph P. Farrell

Farrell delves into the creation of breakaway civilizations by the Nazis in South America and other parts of the world. He discusses the advanced technology that they took with them at the end of the war and the psychological war that they waged for decades on America and NATO. He investigates the secret space programs currently sponsored by the breakaway civilizations and the current militaries in control of planet Earth. Plenty of astounding accounts, documents and speculation on the incredible alternative history of hidden conflicts and secret space programs that began when World War II officially "ended."
292 Pages. 6x9 Paperback. Illustrated. $19.95. Code: BCCW

ORDER FORM

10% Discount When You Order 3 or More Items!

One Adventure Place
P.O. Box 74
Kempton, Illinois 60946
United States of America
Tel.: 815-253-6390 • Fax: 815-253-6300
Email: auphq@frontiernet.net
http://www.adventuresunlimitedpress.com

ORDERING INSTRUCTIONS

- ✓ Remit by USD$ Check, Money Order or Credit Card
- ✓ Visa, Master Card, Discover & AmEx Accepted
- ✓ Paypal Payments Can Be Made To:
  info@wexclub.com
- ✓ Prices May Change Without Notice
- ✓ 10% Discount for 3 or More Items

SHIPPING CHARGES

United States

- ✓ Postal Book Rate { $4.50 First Item / 50¢ Each Additional Item
- ✓ POSTAL BOOK RATE Cannot Be Tracked!
  Not responsible for non-delivery.
- ✓ Priority Mail { $6.00 First Item / $2.00 Each Additional Item
- ✓ UPS { $7.00 First Item / $1.50 Each Additional Item
  NOTE: UPS Delivery Available to Mainland USA Only

Canada

- ✓ Postal Air Mail { $15.00 First Item / $3.00 Each Additional Item
- ✓ Personal Checks or Bank Drafts MUST BE US$ and Drawn on a US Bank
- ✓ Canadian Postal Money Orders OK
- ✓ Payment MUST BE US$

All Other Countries

- ✓ Sorry, No Surface Delivery!
- ✓ Postal Air Mail { $19.00 First Item / $7.00 Each Additional Item
- ✓ Checks and Money Orders MUST BE US$ and Drawn on a US Bank or branch.
- ✓ Paypal Payments Can Be Made in US$ To:
  info@wexclub.com

SPECIAL NOTES

- ✓ RETAILERS: Standard Discounts Available
- ✓ BACKORDERS: We Backorder all Out-of-Stock Items Unless Otherwise Requested
- ✓ PRO FORMA INVOICES: Available on Request
- ✓ DVD Return Policy: Replace defective DVDs only

ORDER ONLINE AT: www.adventuresunlimitedpress.com

10% Discount When You Order 3 or More Items!

Please check: ✓

☐ This is my first order   ☐ I have ordered before

Name	
Address	
City	
State/Province	Postal Code
Country	
Phone: Day	Evening
Fax	Email

Item Code	Item Description	Qty	Total

Please check: ✓

	Subtotal ▶	
	Less Discount-10% for 3 or more items ▶	
☐ Postal-Surface	Balance ▶	
☐ Postal-Air Mail (Priority in USA)	Illinois Residents 6.25% Sales Tax ▶	
	Previous Credit ▶	
☐ UPS (Mainland USA only)	Shipping ▶	
	Total (check/MO in USD$ only) ▶	
☐ Visa/MasterCard/Discover/American Express		

Card Number:

Expiration Date: Security Code:

☐ SEND A CATALOG TO A FRIEND:

---

www.AdventuresUnlimitedPress.com